Reviews of
108 Physiology, Biochemistry and Pharmacology

With 19 Figures and 16 Tables

Springer-Verlag
Berlin Heidelberg GmbH

ISBN 978-3-662-31122-6 ISBN 978-3-540-47824-9 (eBook)
DOI 10.1007/978-3-540-47824-9

Library of Congress-Catalog-Card Number 74-3674

Originally published by Springer-Verlag Berlin Heidelberg New York in 1987
Softcover reprint of the hardcover 1st edition 1987

2127/3130-543210

Contents

Indexed in Current Contents

Rev. Physiol. Biochem. Pharmacol., Vol. 108

Abnormal Expression of Lysosomal Cysteine Proteinases in Muscle Wasting Diseases

NOBUHIKO KATUNUMA and EIKI KOMINAMI

Contents

1 Introduction

Progressive muscle wasting and weakness are two characteristics of various muscle wasting diseases, including muscular dystrophy. In dystrophic muscle there is a net loss of myofibrillar proteins despite an apparent increase in the rate of protein synthesis, suggesting that reduction in muscle protein is largely due to increased protein catabolism.

Consistent with the idea of enhanced protein degradation in dystrophic muscle, there are many reports of increased activities of various types of proteases in this muscle (Pennington and Robinson 1968; Kar and Pearson 1978; Weinstock and Iodice 1969; Iodice et al.

Department of Enzyme Chemistry, Institute for Enzyme Research, School of Medicine, the University of Tokushima, Tokushima 770, Japan

1972; Katunuma et al. 1978; Sanada et al. 1978). Several groups (Bird and Roisen 1986; Katunuma et al. 1983) have suggested that a set of endopeptidases and exopeptidases of the lysosomal apparatus are important in degradation of protein and normal protein turnover in muscle, and that lysosomal proteases are likely to be the proteases involved in increased protein catabolism in dystrophic muscle. However, it is uncertain whether the proteinases involved in the physiological turnover of proteins in normal muscle are also the enzymes responsible for accelerated protein turnover in dystrophic muscle, since secondary changes in muscle, such as concomitant degeneration and regeneration of myofibers, fibroblasts, fat cells, and phagocytes, complicate biochemical analyses of abnormal protein catabolism.

Various combinations of morphological and biochemical approaches to and studies of isolated myoblasts or fibroblasts from patients with Duchenne muscular dystrophy (DMD) have been used to examine this problem. We have characterized lysosomal cysteine proteinases and studied their structures and functions (Katunuma and Kominami 1983). We have also recently investigated the involvement of lysosomes in intramyofibral proteolysis in pathological muscle by immunohistochemical techniques with antibodies against lysosomal cysteine proteinases.

This paper summarizes some of our results on the involvement of the lysosomal system of cathepsins in the pathogenesis of muscle wasting diseases.

2 Structures and Functions of Lysosomal Cysteine Proteinases

2.1 Lysosomal and Nonlysosomal Proteases in Muscle

There are many reports on endopeptidases and exopeptidases in muscle (Pennington 1977; Bird and Carter 1980). However, the most potent endopeptidases are lysosomal cathepsins and cytosolic calcium-activated neutral protease (calpain). Three lysosomal cysteine proteinases, cathepsins B, H, and L, and one aspartate proteinase, cathepsin D, have been demonstrated in cultured muscle cells (Bird et al. 1981; Decker et al. 1980). Since calpain was first reported by Huston and Krebs (1968) and Busch et al. (1972), it has been purified from skeletal muscle, and Ohno et al. (1984) reported the primary structure of calpain from rabbit skeletal muscle. Alkaline serine proteinases that show increased the activity in several protein catabolic conditions, including dystrophy, have also been demonstrated (Katunuma et al. 1975, 1978; Sanada et al.

Table 1. Comparison of properties of cysteine proteinases

	Cathepsin B	Cathepsin H	Cathepsin L
Molecular weight (daltons)	29 000	26 000	30 000
pI	4.9–5.3	6.5–7.5	5.8–6.1
Optimum pH	6.2	6.5	5.5
Catalytic type	Dipeptidyl carboxy-peptidase activity	Aminopeptidase activity	
	Moderate endoprotease activity	Variable endoprotease activity	Very high endoprotease activity
Suitable synthetic substrate	Z-Arg-Arg-NMec	Arg-NMec	Z-Phe-Arg-NMec
Leupeptin inhibition	High	Low	High
Z-Phe-Phe-CHN inhibition	Low	Low	High
E-64 inhibition	High	Low	High
Leu-CH_3Cl inhibition	Low	High	Low

Table 2. Susceptibilities of various myofibrillar proteins to proteinases

	Cathepsin B[a]	Cathepsin H[b]	Cathepsin L[c]
Myosin heavy chain	++	–	+++
L1	–	–	++
L2	–	–	+
L3	–	–	+
Actin	+	–	+
Troponin TN-T	+	+	+++
Troponin TN-I	+	–	++
Troponin TN-C	–	–	–
Tropomyosin	+	+	–
α-Actinin	–	–	+

[a] From Noda et al. (1981)
[b] From Katunuma and Noda (1980)
[c] From Matsukura et al. (1981)

1978; Bird and Carter 1980). However, these enzymes were found to originate from intramuscular mast cells (Woodbury et al. 1978). Recently, Dahlmann et al. (1985) demonstrated a multicatalytic protease in rat skeletal muscle. Besides these endopeptidases, there are also many aminopeptidases, carboxypeptidases, and dipeptidases in the lysosomal and nonlysosomal compartments of muscle (Pennigton 1977; Bird and Carter 1980.

The properties of cathepsins B, H, and L are summarized in Table 1. These proteinases are present in the lysosomes of various tissues and have pH optima of 5.9–6.5 All three have endopeptidase activity; cathepsin L has the strongest activity, followed by cathepsin B. Cathepsins H and B also have exopeptidase activities, cathepsin H having aminopeptidase activity and cathepsin B having carboxydipeptidase activity. The actions of the three proteinases on myofibrillar proteins are shown in Table 2.

Cathepsins B and H from rat liver were the first of the mammalian proteinases to be sequenced (Takio et al. 1983a). The sequence of these enzymes are quite similar to those of papain and actinidine, suggesting that they share a common evolutionary origin with lysosomal cysteine proteinases. We have previously reviewed the structures of lysosomal cysteine proteinases (Katunuma and Kominami 1983).

2.2 Distributions and Localizations of Cathepsins B, H, and L in Rat Tissues and Peripheral Blood Cells

We raised antisera to rat cathepsins B, H, and L, prepared anticathepsins B, H, and L Fab′-horseradish peroxidase conjugates, and measured the amounts of cathepsins in various tissues and peripheral blood cells by enzyme immunoassay (Kominami et al. 1985; Bando et al. 1986). As shown in Table 3, the contents of cathepsins B, H, and L varied greatly between the tissues tested. The concentrations of all three cathepsins were highest in the kidney, and were also high in the liver, spleen, and lung, while skeletal muscle was found to contain quite low levels of all three. The ratio of the levels of the three cathepsins in tissues also varied; for instance, the brain contained much more cathespsins B and L than cathepsin H.

In the kidney, the three cathepsins are all localized in epithelial cells of proximal tubules, but their localizations differ in the liver and brain Staining of cathepsin B was strong in Kupffer cells, and weaker in

Table 3. Level of cathepsins B, H, and L in rat tissues (ng/mg protein) and peripheral blood cells (ng/10^6 cells)

Cells	Cathepsin B	Cathepsin H	Cathepsin L
Kidney	1147±240	1429±360	606±180
Liver	542±122	556±201	120± 52
Spleen	542±122	480±195	137± 61
Lung	151± 26	543±120	115± 38
Brain	310± 86	27± 9	60± 22
Heart	120± 36	82± 22	47± 13
Skeletal muscle	92± 29	52± 16	39± 12
Macrophages			
Resident	194± 28	158± 36	31± 13
Elicited	1259±350	298± 44	90± 40
Lymphocytes	4.1± 1.2	16± 3.2	3.0± 0.9
Neutrophils	6.2± 1.8	16± 4.4	6.1± 2.1

From Kominami et al. (1985) and Bando et al. (1986)

panlobular hepatocytes (Ii et al. 1985); staining for cathepsin H was seen in panlobular hepatocytes, and staining for cathepsin L was strong in centrolobular hepatocytes. Moreover, the distributions of cathepsins in hepatocytes are influenced by nutritional conditions. In meal-fed rats, the three cathepsins were found in the periphery of hepatocytes, possibly in the peribiliary dense bodies, whereas in starved rats stainings for cathepsins L and B, but not cathepsin H, were strong not only in the periphery but also in the perinuclear regions of hepatocytes (Ohshita et al. 1986). The change in distribution of cathepsins in the liver depending on the nutritional state may account for the appearance of many autophagic vacuoles in hepatocytes of starved animals. In brain, cathepsins B and L were found in different types of cells and no cathepsin H was detected in nerve cells.

The three proteinases were not detectable in myofibers from normal rats, but were found in intramuscular macrophages. These results may have been due to low concentrations of these enzymes in normal myofibers, and not to their absence (Kominami et al. 1984a).

Of the peripheral blood cells, macrophages had the highest contents of cathepsins B, H and L, reflecting their potent phagocytic activity (Kominami et al. 1985; Bando et al. 1986). Resident macrophages contained similar levels of cathepsins B and H (160–190 ng/10 cells), but a lower level of cathepsin L. The level of cathepsin B in macrophages obtained from rats after injection of sodium caseinate was 6 times that in resident macrophages, whereas the levels of cathepsins H and L in these elicited macrophages were 2 or 3 times those in resident macrophages. These observations suggest the activation of lysosomal function in inflammatory macrophages and the importance of cathepsin B in lysosomal proteolysis in macrophages.

2.3 Endogenous Inhibitors

We have purified and sequenced two types of cysteine proteinase inhibitors from the cytosol of rat epidermis and liver, respectively (Kominami et al. 1982; Koinami et al. 1984b; Takio et al. 1983b, 1984; Katunuma and Kominami 1986). These two types of inhibitor, cystains α and β, are present in different cells (Kominami et al. 1984b). Their concentrations are low in skeletal muscle and we could not detect them in myofibers using anti-cystatin Fab′-horseradish peroxidase conjugates (Kominami et al. 1984a; Ii et al. 1986). However, an inhibitor (possibly, cystain β) was demonstrated in cultured chick myotubes (Bird and Roisen 1986). Using immunohistochemical methods, we examined the locations of cystains α and β in DMD, and in animal models of muscular dystrophy and distal myopathy with rimmed vacuole formation, but observed no observed no increase of these inhibitors in pathological myofibers. As described later, an abnormal increase in the staining of lysosomal cathepsins was ween in these myofibers. However, endogenous inhibitors have the potential to act as regulatory mechanisms and to mask proteinase activities in muscle cells, and so the levels of both cathepsins and cystatins during the progress of myopathy need to be studied.

3 Mechanism and Regulation of the Autophagic Lysosomal System

Lysosomal cathepsins are packaged within the lyosomal membrane and so cannot act on proteins in the cytoplasm. The mode of action of lysosomal cathepsins on intracellular proteins is shown in Fig. 1.

Intracellular proteins are sequestered in autophagosomes, which eventually fuse with lysosomes containing cathepsins capable of degrading proteins to amino acids and small peptides (Fig. 1.). The first sequestration step is called autophagy. Many studies of autophagy in the

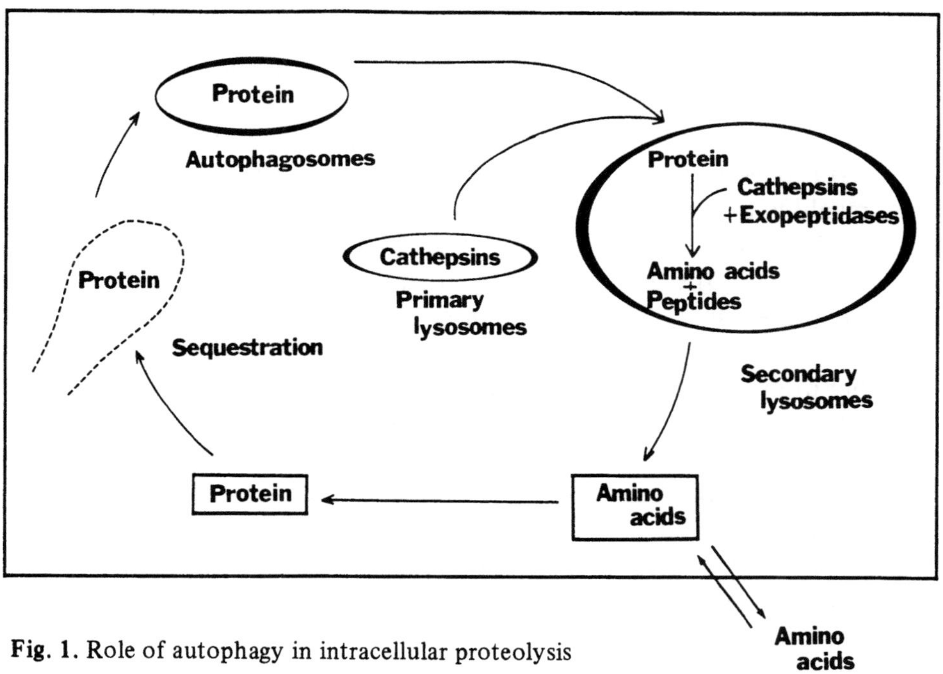

Fig. 1. Role of autophagy in intracellular proteolysis

liver have shown that it is highly responsive to the amino acid demands of cells (Mortimore and Schworer 1977; Poso and Mortimore 1984; Seglen et al. 1980). Lack of amino acids results in enhanced proteolysis, to provide the cells with amino acids for de novo protein synthesis and as a source of energy. However, in muscle, substrate supply does not have such a great influence on proteolysis (Millward et al. 1985), although stimulatory effects of leucine on muscle protein synthesis and inhibitory effects of its keto-acid on proteolysis have been reported (Mitch and Clark 1984).

Another possible function of lysosomes is a homoeostatic function involved in remodeling the enzymic content of the cell, which plays an essential role in cellular adaptation and differentiation. This function is the basis of the heterogeneity of protein turnover. Autophagy is very important in intracellular proteolysis in the liver, accounting for 70%–90% of the total proteolysis. However, it is still uncertain how important lysosomes are in proteolysis in muscle.

Cysteine proteinases play a major role in intralysosomal proteolysis. Administration of leupeptin or E-64, two selective inhibitors of cysteine proteinases, to rats caused the accumulation of numerous autophagic vacuoles in the liver by blocking lysosomal proteolysis (Furuno et al. 1982). However, inhibition of cathepsin D by injection of asialofetuin pepstatin did not result in accumulation of autophagic vacuoles (Furuno et al. 1983).

For determination of the importance of autophagy in maintaining protein homoeostasis, leupeptin was injected and the autophagic vacuoles formed were isolated on Percoll density gradients. In this way highly purified autophagic vacuoles containing less than 1% mitochondrial contaminants were obtained. By examining the uptaken of proteins into autophagic vacuoles after leupeptin administration to starved and meal-fed animals, we showed that accumulation of cytosolic enzymes in autolysosomes is a function of both the rate of autophagosome formation and the concentration of the enzymes in the cytoplasm (Kominami et al. 1983). Furthermore, we observed that the enhanced proteolysis during starvation was due to accelerated sequestration of a number of cytosolic proteins (Ohshita et al. 1986). The levels of tyrosine aminotransferase and lactic dehydrogenase in meal-fed rats were 10- to 20-fold less than those in starved rats (Table 4). However, meal-feeding had no effect on heterophagy (degradation of exogenous proteins; Table 4). Nutritional shift did not influence the rate of clearance of the endocytosed ligand, asialofetuin from the blood stream, the rate of its accumulation in the liver, or its lysosomal level.

In muscle, there is no direct evidence that myofibrillar proteins are isolated and broken down in autophagic vacuoles. The technique

Table 4. Effects of refeeding after starvation on the levels of endogenous enzymes and exogenous asialofetuin in lysosomes from the livers of leupeptin-treated rats. Rats were killed 1 h after injection of leupeptin, and fluorescein-isothiocynate-labeled asialofetuin and liver lysosomes were purified. Data are normalized against the activities of acid phosphatase. Values are means with standard deviations of three rats[a]

Enzyme or proteins	Starving	Refed
Tyrosine aminotransferase (milliunits/unit)	15.1 ± 1.21	1.27 ± 0.43
Lactic dehydrogenase (units/unit)	3.15 ± 0.27	0.17 ± 0.023
Lactic dehydrogenase[b] (μg/unit)	13.34 ± 1.45	0.49 ± 0.10
Catalase (units/unit)	0.65 ± 0.10	0.02 ± 0.0039
5′-Nucleotidase (units/unit)	0.25 ± 0.034	0.11 ± 0.022
Proteins (mg/unit)	3.0 ± 0.055	1.27 ± 0.13
Asialofetuin (μg/unit)	163 ± 29.8	156 ± 22.6

[a] From Ohshita et al. (1986)
[b] Determined by sandwich enzyme immunoassay

described above for liver cannot be applied to muscle, because of the poor distribution of leupeptin injected into muscle and the difficulty in preparing pure lysosomes from muscle. Insulin, corticosteroids, and thyroid hormones are known to influence the regulation of protein synthesis and proteolysis in muscle (Tawa and Goldberg 1986), but it is uncertain whether they stimulate autophagic sequestration.

4 Lysosomal Activation in Pathological Muscle

4.1 Accelerated Degradation of Intracellular Proteins in Distrophic Muscle

In vivo experiments with dystrophic animals have shown increased protein catabolism in dystrophic muscle (Kitchin and Watts 1973; Rouke 1975; Ballard et al. 1979). However, these results are only of average rates of degradation of proteins in muscles with heterogenous cell components, such as atrophic, necrotic, apparently normal and regenerating muscle fibers, and other nonmuscle constituents, and it is uncertain which muscle fibers show enhanced proteolysis. There are recent reports on protein turnover in DMD fibroblasts. Rodemann and Bayreuther (1986) examined net protein synthesis and degradation of total cellular protein and individual proteins in DMD and normal skin fibroblasts by two-dimensional gel elctrophoresis. They found that in DMD fibroblasts the majority of proteins decrease as the result of decreased synthesis and enhanced degradation. The nature of the modifications that lead to the abnormal degradation of individual proteins in DMD fibroblasts and the extent of involvement of lysosomes in the breakdown of abnormal proteins are unknown.

4.2 Increased Proteinase Activities in Pathological Muscle

There are many reports on increases of lysosomal proteinase activities in diseased human and animal muscles. Pearson and Kar (1979) and Katunuma et al. (1978) detected a several-fold increase in cathepsin B activity in diseased human muscles including those with muscular dystrophy, and polymyositis.

We measured the activities of cathepsins B and L in muscle homogenates of 7-week-old normal and dystrophic hamsters (BIO 14.6), and found that they were markedly increased in the white and red muscle and diaphragm of dystrophic animals. However, great variation in the amount of increase of cathepsin B and L activities was observed (Kominami et al. 1984a). These results indicate simply that muscle homogenates consisting of various cell types contain a high activity of cathepsin B. Then, to determine the immunohistochemical localization of cathepsin B in muscle from dystrophic hamsters, we used anti-cathepsin B Fab′-horseradish peroxidase conjugates (Kominami et al. 1984a). Macrophages in or near the myofibers were found to stain very strongly. Necrotic myofibers with myophagocytosis showed positive staining not only for cathepsin B, but also for the other two cathepsins. These results were confirmed in muscles from animal models of muscular dystrophy (dystrophic hamsters and *mdx* mice) and patients with DMD. Since macrophages contain high levels of cathepsins B and H, as shown in Table 4, it seems reasonable that the strong positive staining seen was in macrophages invading muscle fibers.

4.3 Activation of the Autophagic-Lysosomal System in Pathological Muscle

In muscular dystrophy (including dystrophic animals and DMD) the muscle consists of different myofibers undergoing different forms of breakdown, e.g., atrophic and necrotic fibers (very variable necrosis is observed morphologically).

When the soleus muscle from 6-week-old *mdx* mice was subjected to immunohistochemical staining with anticathepsin Fab′-horseradish peroxidase conjugates, strong positive reactions for cathepsin L and B were found in the subsarcolemnal, perinuclear, and possibly intermyofibrillar regions of most atrophic myofibers. Myofibers that showed no apparent atrophy or necrosis were scarcely stained. In skeletal muscle from patients with DMD, stainings for cathepsin B and H were observed in myofibers of some hyline fibers (Fig. 2). Nonnecrotic myofibers surrounding necrotic myofibers did not stain for cathepsin B. These findings suggest that autodigestion, including degradation of damaged myofibrils, by intramyofibral lysosomal proteinases occurs in these pathological muscles (Kominami, Ii and Katunuma 1987).

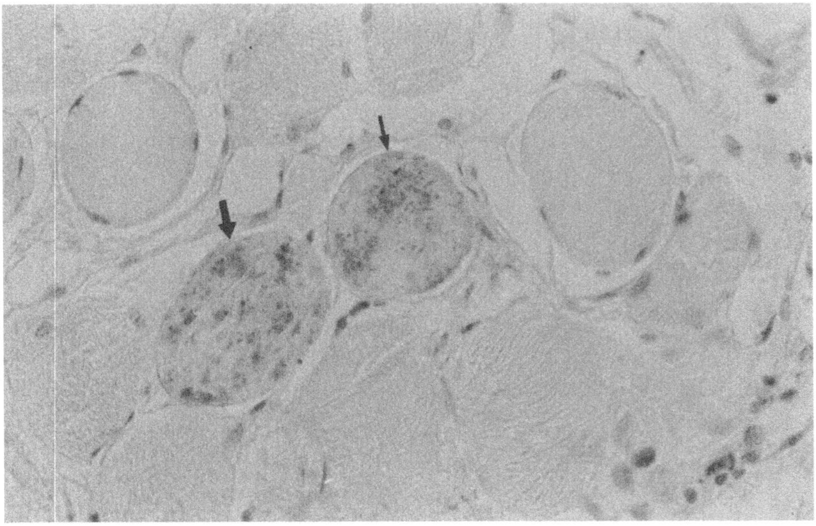

Fig. 2. Localization of cathepsin B in Duchenne muscular dystrophic muscle. Strong positive reactions for cathepsin B are seen in the cytoplasm of many macrophages infiltrating necrotic myofibers (*thick arrow*). Fine granular staining for cathepsin B is seen in hyaline fibers (*thin arrow*). (Immunohistochemical staining with human cathepsin Fab′-horseradish peroxidase conjugates; X 660

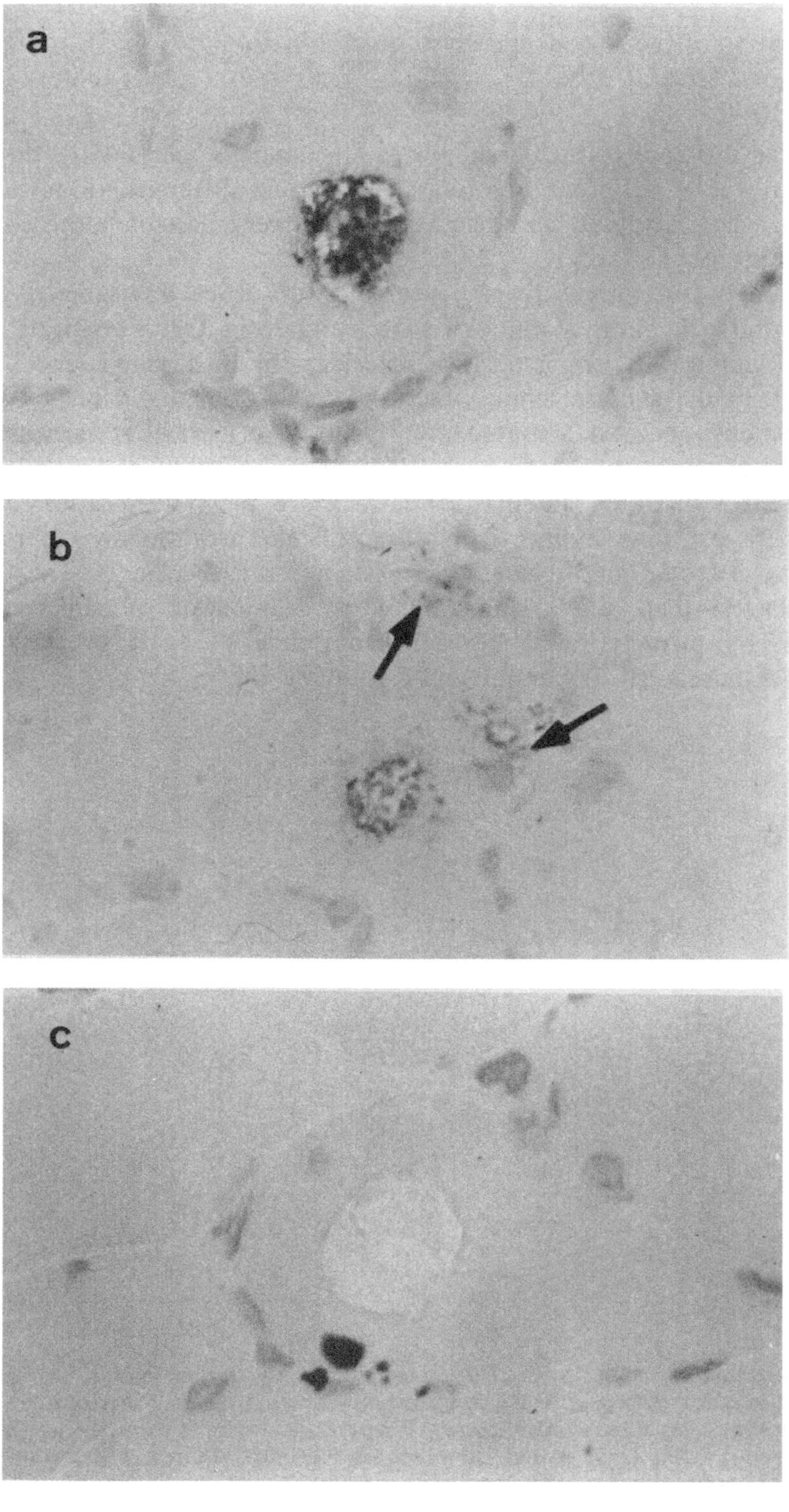
a
b
c

Acute myofiber necrosis can be produced by bupivacaine. Injection of plasmocid into the soleus muscle led to rapid, uniform degeneration of the muscle, followed by dramatic regeneration. Infiltrating macrophages appeared within 6 h and reached a maximum 2 days after the injection (Kominami and Katunuma 1983; Ishiura et al. 1984). When soleus muscle was isolated at a very early stage of myopathy (2 h after injection of plasmocid), the myofibers at the injection sites showed necrosis with edema. The myofibers far from the injection site appeared normal morphologically and histochemically. The fibers in regions between apparently normal and necrotic myofibers could be sublethally injured fibers. Immunohistochemical staining of the soleus showed the presence of cathepsin L in sublethally injured fibers but not in necrotic or apparently normal fibers. This may be evidence of autophagic activity in the fibers before their necrosis. Activation of the intramyofibral lysosomal system was also seen in distal myopathy with rimmed bacuole formation. This disease is an autosomal recessive myopathy involving degeneration of myofibers and the formation of rimmed vacuoles. In this myopathy there was no apparent macrophage infiltration, unlike muscular dystrophy. We showed (Ii et al. 1986) that cathepsins B and H are located at the site of rimmed vacuoles and in some regions within the myofibers (Fig 3.) The granular materials and the rim of the vacuole were shown ultrastructurally to contain autophagic vacuoles with focal destruction of myofilaments. The locations of cathepsins B and H at the site of vacuoles strongly supports the idea that the vacuoles are of autophagic origin.

Long-term administration of chloroquine to rats caused muscle degeneration with many autophagic vacuoles (Sano 1985). At the site of vacuoles in the early stage of myopathy, abnormal increases in cathepsins B and H were demonstrated immunologically by Higuchi et al. (1986) They examined the preventive and therapeutic effect of a cysteine proteinase inhibitor, E-64-d on the experimental chloroquine myopathy. One group of rats was given chloroquine (50 mg kg^{-1}, day^{-1}, intraperitoneally) every day. Another group was given E-64-d (10 mg kg^{-1} day^{-1}, in standard laboratory chow) as well as injections of chloroquine. After 8 weeks the sloeus muscles were isolated. As shown in Table 5, the increase in lyosomal enzyme activities, especially cathepsins B and L,

---

Fig. 3 a–c. Immunohistochemical staining for cathepsins H and B at the sites of rimmed vacuoles. Strong positive staining reactions for cathepsin H are seen within and in the rim of a large vacuole (*a*) and for cathepsin B in smaller vacuoles (*b*) and in subsarcolemmeal regions (*arrows*). Absence of staining at the sites of vacuoles is seen with nonimmune IgG-horseradish peroxidase conjugate (*c*) (Immunohistochemical staining with human cathepsin B or H Fab′-horseradish peroxidase conjugates; X 660)

Table 5. Preventive effect of E-64-d on experimental chloroquine myopathy

	Control	Chloroquine	Chloroquine + E-64-d
Cathepsins B and L (units/μg protein)	0.03± 0.007	0.14± 0.02	0.04± 0.02
Cathepsin H (units/mg protein)	0.10± 0.003	0.43± 0.06	0.20± 0.10
α-Glucosidase (pH 4.01) (units/μg protein)	1.18± 0.10	3.54± 0.26	2.25± 0.54
Acid phosphatase (units/mg protein)	0.08± 0.07	0.22± 0.02	0.11± 0.0
Wet weight of right soleus muscle (mg)	164.0±14.4	57.6 ±10.6	116.6 ±19.1

Modified from Higuchi et al. (1986)

and loss of muscle weight seen in chloroquine myopathy was suppressed markedly by administration of E-64-d. Prevention of muscle degeneration by E-64-d was also confirmed by morphological examination. These results suggest that increase in myofibral cysteine proteinases is important in muscle degeneration in experimental chloroquine myopathy, and that E-64-d may be useful for treatment of myopathy that involves activation of the autophagic-lysosomal system in myofibers.

These results all suggest that activation of the autophagic-lysosomal system can be observed in various types of injured myofiber when the cells are living, and that it is a basic cellular reaction to injury. When the cells are dead, they autolyze. Autolysis is the major event in necrosis. We previously studied the mechanism of postmortem autolysis of sekeletal muscle (Takeichi et al. 1984). In this study, various protase inhibitors, pepstatin, Ep-475 and chymostatin, were injected into the femoral artery of rats before their death, while controls rats received an injection of saline. Then, the postmortem autolysis of myofibrillar proteins with time was examined by SDS-polyacrylamide gel electrophoresis. Tropomyosin was degraded most rapidly, followed by the heavy chain of myosin; actin and tropin T were degraded slowly and were still almost unchanged 2 weeks after death. The degradation was not inhibited by pepstatin, but was inhibited stronlgy by Ep-475 and very strongly by chymostatin. Results suggested that proteases with some specificities, such as lysosomal cysteine proteases and mast cell chymase, are involved in autolysis of muscle after death. Although it is uncertain whether postmortem autolysis occur in the same way, it is likely that autolysis is mediated by both lysosomal and nonlysosomal proteases. Dead cells are disposed of by phagocytes, mainly macrophages.

5 Conclusions and Perspectives

DMD is an X-linked recessive genetic disorder in which the biochemical defect is not yet known. Active studies are in progress to isolate the gene responsible for DMD and/or Becker muscular dystrophy by gene diagnosis with cloned segments. Recently, two cloned segments of human X-chromosome DNA that detect structural alternations within or near the genetic locus responsible for the disorder have been described (Monaco et al. 1985; Kunkel et al. 1986).

On the other hand, much information on the mechanism causing muscle degeneration is also accumulating. Several studies have reported increased levels of compounds associated with free-radical damage in dystrophic muscle in humans (Kar and Pearson 1979) and chickens (Mizuno 1984). In addition, the levels of antioxidant enzymes are reported to be increased in dystrophic muscle of humans (Kar and Pearson 1979; Matkovicks et al. 1982) and chickens (Mizuno 1984), a change that is associated with oxidative stress in other systems (Denke and Fanburg 1980). The change in antioxidant enzyme activity, and the presence of by-products of free-radical damage, both support the idea that free radicals are involved in the pathogenesis of inherited muscular dystrophy. Oxygen radicals have been implicated as a cause of cellular death (McCord 1974). Various oxygen species (O_2^-, H_2O_2, $\cdot OH$, and O_2^-) are highly reactive and can alter most types of cellular macromolecules. In vitro, they have been shown to oxidize proteins, peroxidize unsaturated fatty acid, damage nucleic acids, and cleave polysaccharides.

Another mediator of cellular injury is Ca^{2+}. In the muscle of DMD patients intracellular Ca^{2+} is increased (Bodensteiner and Engel 1978), and many abnormalities observed in DMD muscle in vivo, such as enhanced protein degradation, could be due to increased intracellular Ca^{2+}, Fingerman et al. (1984) suggested that Ca^{2+} influx is important in the regulation of protein synthesis and degradation in DMD fibroblasts in vitro. An increase of Ca^{2+} can cause contraction of myofibrils and disassembly of myofilaments, and may also activate calpain, although activation of calpain in dystrophic muscle has not been shown directly. Increased Ca^{2+} influx is also suggested to activate lysosomal proteolysis (Goldberg et al. 1984). Thus, the myocellular injury seen in DMD and other muscle wasting diseases could be mediated by free radicals and/or increased Ca^{2+} influx. The autophagy seen in various types of injured myofibers may be a cellular response to remove abnormal proteins produced by injury and to maintain homoeostasis. However, the question remains of what triggers activation of the autophagic lysosomal system in injured muscle. And is disassembly of myofilaments by extralysosomal

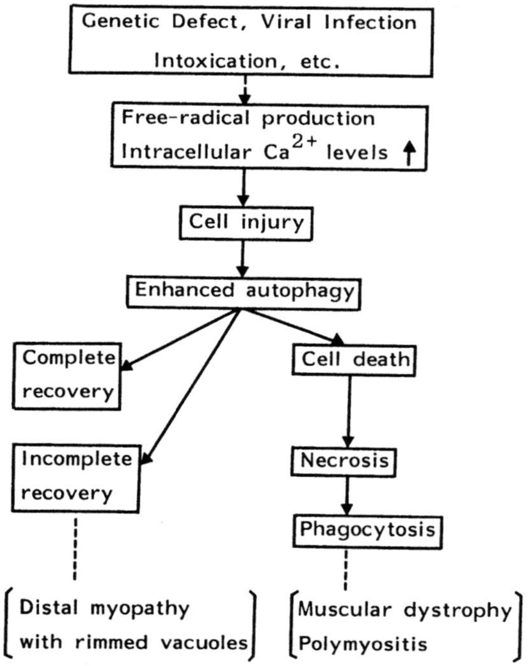

Fig. 4. Hypothesis of pathogenesis of muscle wasting diseases

proteases such as calpain a prerequisite for lysosomal proteolysis? There is some evidence that myofibrillar proteins are broken down in lysosomes, but further studies of autophagy and lysosomal involvement in proteolysis in normal and pathological muscles are required. The cell may recover if the injurious influence is removed. This autophagic homoeostatic action may be incomplete in distal myopathy with rimmed vacuoles. The prolonged presence of rimmed vacuoles containing autophagic vacuoles with undigested materials and lysosomal cathepsins suggests the low digestibility of lysosomes and the subsequent slow regression of the autophagic vacuoles. If severe cellular injury is maintained, autophagy may be followed by cellular death and necrosis, when normal intracellular reactions such as autophagy cease. In muscle, necrosis is followed by phagocytosis by inflammatory cells, chiefly macrophages. Figure 4 shows our hypothesis of the pathogenesis of muscle wasting diseases.

Acknowledgments. The authors are indebted to Ms. E. Inai for assistance in preparation of the manuscript. Original work from this laboratory reported in the text was supported by grants from the Ministry of Education, Science and Culture of Japan, and from the National Center for Nervous, Mental and Muscular Disorders of the Ministry of Health and Welfare of Japan.

References

Ballard FJ, Thomas FM, Sterun LM (1979) Increased turnover of muscle contractile proteins in Duchenne muscular dystrophy as assessed by 3-methylhistidine and creatinine excretion. Clin Sci 56:347–352

Bando Y, Kominami E, Katunuma N (1986) Purification and tissue distribution of rat cathepsin L. J Biochem (Tokyo) 100:35–42

Bird JWC, Carter JH (1980) Proteolytic enzymes in striated and non-striated muscle. In: Wildenthal K (ed) Degradative processes in heart and skeletal muscle. Elsevier, Amsterdam, pp 51–85

Bird JWC, Roisen FJ (1986) Lysosomes in muscle: developmental aspects, enzyme activities, and role in protein turnover. In: Engel AG, Banker BQ (eds) Myology. McGraw-Hill, New York, pp 745–767

Bird JWC, Roisen FJ, Yorke G, Lee JA, McElligott MA, Triemer DJ, St. John A (1981) Lysosomes and proteolytic activities in cultured striated muscle cells. J Histochem Cytochem 29:431–439

Bodensteiner JB, Engel AG (1978) Intracellular accumulation in Duchenne dystrophy and other myopathies: a study of 567,000 muscle fibers in 114 biopsies. Neurology (Minneapolis) 28:439–448

Busch WA, Stromer MH, Goll DE, Suzuki A (1972) Ca^{2+} specific removal of Z-line from rabbit skeletal muscle. J Cell Biol 52:367–381

Dahlmann B, Kuehn L, Rutschmann M, Reinauer H (1985) Purification and characterization of a multicatalytic high-molecular-mass proteinase from rat skeletal muscle. Biochem J 228:161–170

Decker RS, Decker M, Pool AR (1980) The distribution of lysosomal cathepsin D in cardiac myocytes. J Hitochem Cytochem 28:231–237

Denke SM, Fanburg JL (1980) Normobasic oxygen toxicity of the lung. N Engl J Med 303:76–86

Fingerman E, Campisi J, Pardee AB (1984) Defective Ca^{2+} metabolism in Duchenne muscular dystrophy: effects on cellular and viral growth. Proc Natl Acad Sci USA 81:7617–7621

Fong D, Calhoun DH, Hsieh WT, Lee B, Wells RT (1986) Isolation of a c-DNA clone for the human lysosomal proteinase cathepsin B. Proc Natl Acad Sci USA 83:2909–2913

Furuno K, Ishikawa T, Kato K (1982) Appearance of autolysosomes in rat liver after leupeptin treatment. J Biochem (Tokyo) 91:1485–1494

Furuno K, Miwa N, Kato K (1983) Receptor-mediated introduction of pepstatin-asialofetuin conjugate into lysosomes of rat hepatocytes. J Biochem (Tokyo) 93:249–256

Gelard KW, Schneider DL (1979) Evidence for degradation of myofibrillar proteins in lysosomes. J Biol Chem 254:11798–11805

Goldberg AL, Baracos V, Rodemann HP, Dianarello CA, Waxman L (1984) Control of protein degradation in muscle by prostaglandins, Ca^{2+}, and leukocyte pyrogen (interleukin 1). Fed Proc 43:1301–1306

Goldspink DF, Goldspink G (1977) Age related changes in protein turnover and ribonucleic acid of diaphragm muscle of normal and dystrophic hamsters. Biochem J 162:191–194

Higuchi I, Ishiura S, Sugita H, Ii K, Katunuma N (1986) The therapeutic effect of cysteine proteinase inhibitor. EST, on experimental chloroquine myopathy. (In Japanese, abstract) English. Clin Neurol 26:928–936

Huston RB, Krebs EG (1968) Activation of skeletal muscle phosphorylase kinase by Ca^{2+}. II. Identification of the kinase activating factor as a proteolytic enzyme. Biochemistry 7:2116–2122

Ii K, Hizawa K, Kominami E, Bando Y, Katunuma N (1985) Different immunolocalization of cathepsins B, H and L in the liver. J Histochem Cytochem 33: 1173–1175

Ii K, Hizawa K, Nonaka L, Sugita H, Kominami E, Katunuma N (1986) Abnormal increases of lysosomal cysteine proteinases in rimmed vacuoles in the skeletal muscle. Am J Pathol 122:193–198

Iodice AA, Clin J, Perker S, Weinstock IM (1972) Cathepsins A, B, C D and autolysis during development of breast muscle of normal and dystrophic chickens. Arch Biochem Biophys 152:166–174

Ishiura S, Nonaka I, Nakase H, Tada A, Sugita H (1984) Two step mechanism of myofibrillar protein degradation in acute plasmocid-induced muscle necrosis. Biochim Biophys Acta 798:333–342

Jackson MJ, Johnes DA, Edwards RHT (1984) Techniques for studying radical damage in muscular dystrophy. Med Biol 62:135–138

Kar NC, Pearson CM (1978) Dipeptidyl peptidases in human muscle diseases. Clin Chim Acta 82:185–192

Kar NC, Pearson CM (1979b) Catalase, superoxide dismutase, glutathione reductase and thiobarbituric acid reactive products in normal and dystrophic human muscle. Clin Chim Acta 94:277–280

Katunuma N, Kominami E (1983) Structures and functions of lysosomal thiol proteinases and their endogenous inhibitors. In: Horecker BL, Stadtman ER (eds) Current topics in cellular regulation, vol 22, Academic, New York, pp 71–101

Katunuma N, Kominami E (1986) Molecular basis of intracellular regulation of thiol proteinase inhibitors. In: Horecker BL, Stadtman ER (eds) Current topics in cellular regulation, vol 27. Academic, New York, pp 345–360

Katunuma N, Noda T (1980) Cathepsin B, H and L and intracellular protein degradation in atrophic muscle of muscular dystrophy. In: Ebashi S (ed) Muscular dystrophy. University of Tokyo Press, Tokyo, pp 225–237

Katunuma N, Kominami E, Kobayashi K, Banno Y, Suzuki K, Chichibu K, Hamaguchi Y, Katsunuma T (1975) Studies on new intracellular proteases in various organs of rat I. Purification and comparison of their properties. Eur J Biochem 52:37–50

Katunuma N, Yasogawa N, Kito K, Sanada Y, Miyoshi K (1978) Abnormal expression of a serine protease in human dystrophic muscle. J Biochem (Tokyo) 83:625–628

Katunuma N, Kominami E, Noda T, Isogai K (1983) Lysosomal thiol proteinases and muscular dystrophy. In: Ebashi S, Ozawa E (eds) Muscular dystrophy: biomedical aspects. Springer, Berlin Heidelberg New York Tokyo, pp 247–256

Kitchin SE, Watts DC (1973) Comparison of the turnover patterns of total and individual muscle proteins in normal mice and those with hereditary muscular dystrophy. Biochem J 136:1017–1028

Kominami E, Katunuma N (1983) Lysosomal thiol proteinases in plasmocid-induced myopathy. In: Annual report of clinical research group of muscular dystrophy, Japan, pp 195–197 (In Japanese, author's translation)

Kominami E, Wakamatsu N, Katunuma N (1982) Purification and characterization of thiol proteinase inhibitor from rat liver. J Biol Chem 257:14648–14652

Kominami E, Hashida S, Kahirallah EA, Katunuma N (1983) Sequestration of cytoplasmic enzymes in an autophagic vacuole-lysosomal system induced by injection of leupeptin. J Biol Chem 258:6093–6100

Kominami E, Bando Y, Ii K, Hizawa K, Katunuma N (1984a) Increases in cathepsin B and L thiol proteinase inhibitors in muscle of dystrophic hamsters. Their localization in invading phagocytes. J Biochem (Tokyo) 96:1941–1948

Kominami E, Bando Y, Wakamatsu N, Katnunuma B (1984b) Different tissue distributions of two types of thiol proteinase inhibitors from rat liver and epidermis. J Biochem (Tokyo) 96:1437–1442

Kominami E, Tsukahara T, Bando Y, Katunuma N (1985) Distribution of cathepsins B and H in rat tissues and peripheral blood cells. J Biochem (Tokyo) 98:87–93
Kominami E, Ii K and Katunuma N (1987) Activation of the intramyofibral autophagic lysosomal system in muscular dystrophy. Am J Pathol 127(3):21–26
Kunkel LM et al. (1986) Analysis of deletions in DNA from patients with Becker and Duchenne muscular dystrophy. Nature 322:73–77
Matkovicks B, Laszlo A, Szabo L (1982) A comparative study of superoxide dismutase, catalase and lipid peroxidation in red blood cells from muscular dystrophy patients and normal controls. Clin Chim Acta 118:289–292
Matsukura U, Okitani A, Nishimoto T, Kato H (1981) Mode of degradation of myofibrillar proteins by endogenous protease, cathepsin L. Biochim Biophys Acta 662;41–47
McCord JM (1974) Free radicals and inflammation: protection of synovial fluid by superoxide dismutase. Science 195:529–531
Millward D, Bates P, Brown JG, Cox M, Jepson M, Pell J (1985) In: Khairallah E, Bond JS, Bird JWC (eds) Intracellular protein catabolism. Liss, New York, pp 531–542
Mitch WE, Clark AS (1984) Specificity of the effects of leucine and its metabolism on protein degradation in sekeletal muscle. Biochem J 222:579–586
Mizuno Y (1984) Changes in superoxide dismutase, catalase, glutathione peroxidase, and glutathione reductase activities and thiobarbituric reactive product levels in early stage of development in dystrophic chickens. Exp Neurol 84:58–73
Mortimore GE, Schworer CM (1977) Induction of autophagy by amino acid depriation in perfused rat liver. Nature 270:174–176
Monaco AP, Bertelson CJ, Middlesworth W, Colletti CA, Aldridge J, Fischbeck KH, Barlett R, Pericak-Vance MA, Roses AD, Kunkel LM (1985) Detection of deletions spanning the Duchenne muscular dystrophy locus using a tightly linked DNA segment. Nature 316:842–845
Noda T, Isogai K, Hayashi H, Katunuma N (1981) Susceptibilities of various myofibrillar proteins to cathepsin B and morphological alteration of isolated myofibrils by this enzyme. J Biochem (Tokyo) 90:371–379
Ohno S, Emori Y, Imajoh S, Kawasaki H, Kisaragi M, Suzuki K (1984) Evolutionary origin of a calcium dependent protease by fusion of genes for a thiol protease and a calcium-binding protein? Nature 313:566–570
Ohshita T, Kominami E, Ii K, Katunuma N (1986) Effect of starvation and refeeding on autophagy and heterophagy in rat liver. J Biochem (Tokyo) 100:623–632
Pearson CH, Kar NC (1979) Muscle breakdown and lysosomal activation. (Biochemistry). Ann NY Acad Sci 317:465–477
Pennington RJ (1977) Proteinases in muscle. In: Barrett AJ (ed) Proteinases in mammalian cells and tissues. Elsevier, Amsterdam, pp 515–543
Pennington RJ, Robinson CM (1968) Cathepsin activity in normal and dystrophic human muscle. Enzymol Biol Clin 9:175–182
Poso AR, Mortimore EG (1984) Requirement for alanine in the amino acid control of deprivation-induced protein degradation in liver. Proc Natl Acad Sci USA 81:4270–4274
Rodemann HP, Bayreuther K (1986) Differential degradation of (^{35}S)-methionine polypeptides in Duchenne muscular dystrophy skin fibroblasts in vitro. Proc Natl Acad Sci USA 83:2086–2090
Rouke AW (1975) Myosin in developing normal and dystrophic chicken pectoralis. I. Synthesis and degradation. J Cell Physiol 86:343–352
Sanada Y, Yasogawa N, Katunuma N (1978) Serine protese in mice with hereditary muscular dystrophy. J Biochem (Tokyo) 83:27–33
Sano M (1985) Biochemical studies in experimental chloroquine myopathy (In Japanese, English abstract). Clin Neurol 25:627–636
Seglen PO, Gordon PB, Poli A (1980) Amino acid inhibition of the autophagic/lysosomal pathway of protein degradation in isolated rat hepatocytes. Biochem Biophys Acta 630:103–118

Segundo BS, Chan SJ, Steiner DF (1985) Identification of cDNA clones encoding a precursor of rat liver cathepsin B. Proc Natl Acad Sci USA 82:2320–2324

Takeichi S, Tokunaga I, Yoshima K, Maeiwa M, Bando Y, Kominami E, Katunuma N (1984) Mechanisms of postmortem autolysis of sekeletal muscle. Biochem Med 32:341–348

Takio K, Towatari T, Katunuma N, Teller DC, Titani K (1983a) Homology of amino acid sequences of rat liver cathepsins B and H with that of papain. Proc Natl Acad Sci USA 80:3666–3670

Takio K, Kominami E, Wakamatsu N, Katunuma N, Titani K (1983b) Amino acid sequence of rat liver thiol proteinase inhibitor. Biochem Biophys Res Commun 115:902–908

Takio K, Kominami E, Bando Y, Katunuma N, Titani K (1984) Amino acid sequence of rat epidermis thiol proteinase inhibitor. Biochem Biophys Res Commun 121:149–154

Tawa NE Jr, Goldberg AL (1986) Protein and amino acid metabolism in muscle. In: Engel AG, Banker BQ (eds) Myology. McGraw-Hill, New York, pp 721–743

Weinstock IM, Iodice AA (1969) Acid hydrolase activity in muscular dystrophy and denervation atrophy. In: Dingle JT, Fell FB (eds) Lysosomes in biology and pathology. Elsevier, New York, pp 450–468

Woodbury RG, Everitt M, Sanada Y, Katunuma N, Lagunoff D, Neurath H (1978) A major serine protease in rat sekeletal muscle; evidence for its mast cell origin. Proc Natl Acad Sci USA 75:5311–5313

Rev. Physiol. Biochem. Pharmacol., Vol. 108

Hereditary Disorders of the Urea Cycle in Man: Biochemical and Molecular Approaches

TAKEYORI SAHEKI, KEIKO KOBAYASHI, and ITURO INOUE

Contents

Department of Biochemistry, Faculty of Medicine, Kagoshima University
Kagoshima 890, Japan

1 Introduction

The urea cycle is a fundamental mechanism in mammals for detoxifying waste nitrogen. It involves five enzymes – carbamylphosphate synthetase (CPS), ornithine transcarbamylase (OTC), argininosuccinate synthetase (ASS), argininosuccinate lyase (ASL), and arginase – and protein which transports ornithine from the cytosol to the mitochondria. *N*-Acetylglutamate synthetase is another enzyme which is essential for the proper operation of the cycle. Other factors necessary for urea synthesis are the electron transport system which generates ATP, the TCA cycle which supplies aspartate, and the dibasic amino acid transport system at the cell membrane of the intestine and kidney which supplies ornithine, a carrier of nitrogen. Malfunction of any of these varieties of enzymes or proteins generally results in hyperammonemia.

The biochemistry of urea cycle enzymes and the clinical features of hereditary disorders have been reviewed in *The Urea Cycle* (Grisolia et al. 1976) and *Metabolic Basis of Inherited Diseases* (Stanbury et al. 1978). In this review, we shall describe recent advances in biochemical and molecular research into hereditary disorders of the urea cycle.

2 Hyperammonemia

2.1 Causes

Table 1 lists the causes of hyperammonemia. Besides deficiencies of the urea cycle enzymes, hyperammonemia is caused by hereditary disorders of membrane transport proteins such as the mitochondrial ornithine transport system and the dibasic amino acid transport system, which is located at the renal and intestinal basolateral membrane (Desjeux et al. 1980). Both are thought to cause hyperammonemia owing to a reduced level of ornithine, a nitrogen carrier, necessary for OTC reaction.

Table 1. Causes of hyperammonemia

Hereditary

1. Deficiencies of urea cycle enzymes
   a) Congenital hyperammonemia type I
   deficiency of carbamylphosphate synthetase
   b) Congenital hyperammonemia type II
   deficiency of ornithine transcarbamylase
   c) Citrullinemia
   deficiency of argininosuccinate synthetase
   d) Argininosuccinic aciduria
   deficiency of argininosuccinate lyase
   e) Argininemia
   deficiency of arginase
   f) *N*-Acetylglutamate synthetase deficiency
2. Deficiencies of membrane transport proteins
   a) Hyperornithinemia, hyperammonemia, and homocitrullinuria
   b) Lysinuric protein intolerance
3. Hyperammonemia secondary to other hereditary metabolic disorders
   a) Organic acidemia
   propionic acidemia
   isovaleric acidemia
   methylmalonic acidemia
   b) Others
   deficiency of pyruvate carboxylase
   deficiency of pyruvate decarboxylase
   Zellweger syndrome

Nonhereditary

1. Neonatal transient hyperammonemia
2. Reye syndrome
3. Portacaval shunt and liver damage
4. Hyperammonemia due to administration of chemical reagents such as hypoglycin, 4-pentenoic acid, and valproic acid

The pathogenesis may be very similar in both hyperammonemia caused by organic acidemia and that caused by the administration of chemical reagents such as hypoglycin and 4-pentenoic acid, which cause Jamaican vomiting sickness, and also in that caused by valproic acid, which is an antiepileptic drug. Blood ammonia levels in organic acidemia have been shown to be positively correlated to the concentration of organic acids accumulated in the blood (Coude et al. 1982). The mechanism of hyperammonemia seems to be as follows: the inhibition of CPS activity is caused by a decreased hepatic level of *N*-acetylglutamate which results from a decreased level of acetyl-CoA (Coude et al. 1983). However, other possibilities should not be neglected (Gruskay and Rosenberg 1979; Coude et al. 1979; Cathelineau et al. 1979). A similar situation is produced by the administration of hypoglycin, 4-pentenoic acid, and valproic acid. These reagents, after being converted to their CoA derivatives, inhibit fatty acid breakdown (Corredor et al. 1967; Holland and Sherratt 1973), resulting in a decrease of acetyl-CoA.

Igarashi et al. (1986) reported a case of Zellweger syndrome accompanying hyperammonemia. Zellweger syndrome is an autosomal recessive hereditary disorder and has been found to include a deficiency in peroxisomal enzymes of β-oxidation (Goldfischer et al. 1973; Suzuki et al. 1986), which may cause the accumulation of very long chain fatty acids. The mechanism for hyperammonemia in the Zellweger syndrome is not known, but may be similar to those stated above for organic acidemia.

Coude et al. (1981) and Pollock et al. (1986) reported four and two neonates respectively with pyruvate carboxylase deficiency which showed not only lactic acidosis but also hyperammonemia and an elevated level of plasma citrulline. They postulated a decreased supply of oxaloacetate and then aspartate as the cause of the dysfunction of the urea cycle. Similarly, hyperammonemia accompanying deficiency of pyruvate decarboxylase has been reported (Matsuo et al. 1982).

Prematurity of the urea cycle enzymes has been postulated as the cause of neonatal transient hyperammonemia (Walser 1983).

2.2 Pathophysiology

Almost all of the clinical symptoms of hyperammonemia, such as vomiting, lethargy, stupor, seizures, and coma, are connected with the central nervous system, and prolonged hyperammonemia results in mental retardation. Under various hyperammonemic states, no specific changes in serum amino acids were observed but there was a remarkable increase of glutamine in the cerebrospinal fluid (Levin et al. 1969; James et al. 1979; Kitajima et al. 1986). These results suggest that ammonia mainly affects the function of the central nervous system.

As shown by the experiments with animal models (Bachmann and Colombo 1983; Mans et al. 1983; Inoue et al., in preparation), high blood ammonia increases the concentration of glutamine, tyrosine, phenylalanine, and tryptophan in the brain. The latter three amino acids are precursors for various neurotransmitters. These considerations suggest that ammonia causes disturbances of neurotransmitter metabolism in the brain. Bachmann and Colombo (1984) reported data confirming increased metabolism of tryptophan in the brain of hyperammonemic mice.

Jonung et al. (1984) described how the administration of methionine sulfoximine, an inhibitor of glutamine synthetase, inhibits the possible pathological effects of ammonia; namely it causes decreases in the concentrations of tyrosine and phenylalanine, as well as glutamine, in the brain. In the case of experimental animals with a portacaval shunt (Rigotti et al. 1985), methionine sulfoximine caused similar changes in brain amino acids without causing significant changes in plasma amino acids such as tyrosine and phenylalanine. These results again confirm that the metabolic effect of ammonia plays an important role in the clinical symptoms of hyperammonemia. The precise mechanism, however, is still not fully understood.

3 Carbamylphosphate Synthetase

3.1 Biochemistry

Carbamylphosphate synthetase (CPS) is the first-step enzyme of the urea cycle. It catalyzes the following reaction:

ammonia + bicarbonate + 2 ATP = carbamylphosphate + 2 ADP + Pi

The CPS operating in the urea cycle is known as CPS I. Two other kinds of CPS have been described: CPS II, an *N*-acetylglutamate-independent enzyme functioning in pyrimidine biosynthesis (which utilizes glutamine), and CPS III, an *N*-acetylglutamate-dependent enzyme which plays a role in the formation of urea with glutamine as a nitrogen source for osmoregulation in marine elasmobranches. CPS I uses ammonia as a nitrogen substrate and requires *N*-acetylglutamate. In this paper, unless otherwise stated, CPS refers to CPS I.

Guthöhrlein and Knappe (1968) purified CPS from rat liver by stabilizing the enzyme with glycerol, mercaptoethanol, and KCN, and the enzyme has been purified from rat liver (Raijman and Jones 1976; Lusty 1978; Powers 1981) and human liver (Pierson and Brien 1980; Rubio and Grisolia 1981) using various procedures. Purified CPS is stable in the presence of glycerol, dithiothreitol, and ammonium sulfate at $-20^{\circ}C$ (Raijman and Jones 1976; Lusty 1978) (human CPS was also stable under the same conditions in our experiment), and in the presence of ATP-Mg and dithiothreitol in Tris-acetate pH 8.0 at 4°C (Pierson and Brien 1980). Purified human CPS is a mixture of monomers and dimers (up to 90% of the native enzyme is monomers) with a subunit molecular weight of 160 000 daltons (Rubio and Grisolia 1981) or 165 000 daltons (Pierson and Brien 1980) and with kinetic parameters as shown in Table 2. Non-Michaelis-Menten kinetics are not observed with either of the substrates and the activator.

Immunological tests revealed very extensive similarities between CPS in man and in rat (Rubio and Grisolia 1981); difference of reactivity between different classes of CPS with anti-rat-CPS antiserum was detected only by microcomplement fixation test. Structural similarities were shown by immunoblot assay among CPS I, CPS II, and CPS III (Devaney and Powers-Lee 1984). CPS II protein from *E. coli* also reacted with anti-rat-CPS antiserum. These results indicate structural conservation among the various classes of CPS and suggest evolutionary construction of the enzyme.

Three research groups independently isolated cDNA clones for rat and human CPS in 1984 (Broglie and Lusty, unpublished results described by Ryall et al. 1984; Adcock and O'Brien 1984; de Groot et al.

Table 2. Physicochemical properties and kinetic parameters of purified human carbamylphosphate synthetase

Parameters	Pierson and Brien (1980)	Rubio and Grisolia (1981)
Molecular weight		
subunit	165 000	160 000±5000
native enzyme	178 000	190 000
pI	6.05	6.23–6.26
Optimal pH	7.8	7.8–8.0
Specific activity[a]	1.5	1.7
Km values (m*M*) for		
ATP	1.1	0.26
HCO_3^-	6.7	2.2
ammonia	0.8	1.3
N-acetylglutamate	0.1	0.15

[a] Expressed as μmoles of carbamylphosphate formed per minute per milligram protein at 37°C

1984) and used them to quantify mRNA levels in the rat liver during dietary changes and development, and in hepatoma cell lines. Adcock and O'Brien (1984) gave restriction maps of human CPS cDNA clones. From Northern blot analysis, the size of CPS mRNA was calculated to be 6 kb by de Groot et al. (1984), 5 kb by Adcock and O'Brien (1984), and 6.5–6.7 kb by Ryall et al. (1984).

3.2 Deficiency

Deficiency of CPS, or congenital hyperammonemia type I, was first described by Freeman et al. (1970). Nowadays this disorder seems not to be as rare as once thought; whereas only six cases were reviewed in 1976 by Shih, 26 cases were summarized in 1983 by Walser. In Japan eight cases of CPS deficiency have been found and analyzed in the past few years (Arashima and Matsuda 1972; Kakinuma et al. 1984; Goto et al. 1983; Saheki et al., unpublished data).

Patients with CPS deficiency show the typical symptoms of hyperammonemia – poor feeding, recurrent vomiting, lethargy, stupor, and seizures. There are no specific changes in serum amino acids other than nonspecific increases in glutamine and alanine and no increased excretion of orotic acid. Similar clinical presentations have been described for the deficiency of *N*-acetylglutamate synthetase (Bachmann et al. 1981). Differentiation between deficiency of CPS and of *N*-acetylglutamate synthetase may be possible only by determining the activities of both enzymes in the liver, although so far no second case of the latter deficiency has been described.

Enzymatic analysis has revealed two groups, with complete or almost complete absence of CPS activity in one group, to which neonatal-onset patients belong, and partial CPS deficiency in the other group. In Walser's review (1983), 11 of 26 cases belong to the former group. In Japan, except the first case (Arashima and Matsuda 1972), all, including our recent three cases, have been shown to be completely or almost completely lacking in CPS activity.

Absence of CPS protein in the liver of a CPS-deficient patient was first shown by Raijman (1979) using SDS-polyacrylamide gel electrophoresis (SDS-PAGE), which is easy to perform and gives very clear results. Figure 1 shows such an analysis done with a CPS-deficient patient in the author's laboratory. CPS is composed of a polypeptide with a relatively high molecular weight (160 000–165 000 daltons) and is abundant in the liver [1.5% of total liver protein or 0.2 m*M* according to Rubio and Grisolia (1981)], which can be distinguished from other polypeptides on SDS-PAGE.

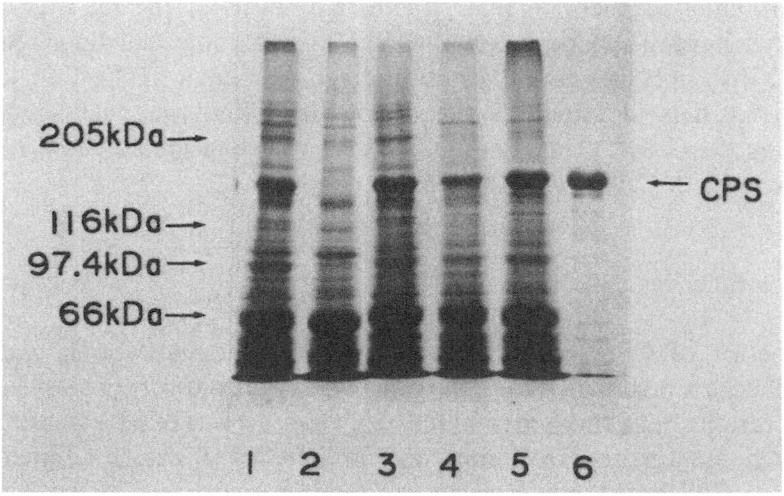

Fig. 1. SDS-polyacrylamide gel electrophoresis of liver homogenates from a patient deficient in carbamylphosphate synthetase (*CPS;* lane 2) and from control subjects (lanes 1, 3, 4, 5), and of partially purified human CPS (lane 6)

Ohtake et al. (1984) reported two sisters with complete deficiency of CPS; no detectable CPS protein and no cross-reactive material (CRM) were found on SDS-PAGE and Ouchterlony double immunodiffusion. Analysis with a cell-free translation system failed to find any translational activity of mRNA coding for CPS with total RNA extracted from the liver of a patient with CPS deficiency, although numerous newly synthesized proteins having molecular weights up to 200 000 daltons were found with RNAs extracted from the patient and control subjects and a presumed precursor form of CPS was detected with RNA from control subjects.

Using cloned cDNA for human CPS and restriction endonuclease techniques, Fearon et al. (1985) performed a genetic analysis of CPS deficiency; they observed no abnormality in the number or size of the hybridizing DNA fragments from six patients with complete deficiency of CPS and one patient with partial deficiency, suggesting that no gross alteration affected the CPS genes. However, they found a frequent restriction fragment length polymorphism (RFLP) at the CPS gene locus using *BqlI* restriction endonuclease and cosegregation of RFLP with the CPS deficiency phenotype, and this will enable prenatal diagnosis of CPS deficiency.

4 Ornithine Transcarbamylase

4.1 Biochemistry

Ornithine transcarbamylase (OTC) is the second enzyme of the urea cycle and is located in the mitochondria. It catalyzes the conversion of carbamylphosphate and ornithine to citrulline and inorganic phosphate. OTC, purified from steer (Marshall and Cohen 1972), rat (Lusty et al. 1979; Hoogenraad et al. 1980), mouse (Briand et al. 1981), and human (Pierson et al. 1977; Kalousek et al. 1978) livers, is a homotrimer with a subunit molecular weight of 36 000–38 000 daltons. The physicochemical properties and kinetic parameters of human OTC are listed in Table 3.

OTC is found almost exclusively in the liver and in the intestine, where the enzyme, present at only about 5% of the liver activity, works on supplying citrulline to the blood. The enzyme is encoded in a nuclear gene, is synthesized in the cytosol as a precursor form with a molecular weight of about 4000 daltons more than the mature subunit, and is transported into the mitochondria releasing a part of the precursor, or leader peptide, as with CPS. Many papers have been published on the mechanism of OTC translocation (Conboy et al. 1979; Mori et al. 1981; Argan and Shore 1985; Horwich et al. 1985).

A cloned cDNA containing C-terminal sequences corresponding to the mRNA encoding rat liver OTC was first isolated by Horwich et al.

Table 3. Physicochemical properties and kinetic parameters of purified human ornithine transcarbamylase

Parameters	Pierson et al. (1977)	Kalousek et al. (1978)
Molecular weight		
subunit	36 500	38 000
native enzyme	110 000	100 000
pI	7.95	6.8±0.1
Optimal pH	8.0	7.7
Specific activity[a]	86.5	233
Km values (m*M*) for		
ornithine	0.2	0.4
carbamylphosphate	0.09	0.16

[a] μmoles citrulline formed per minute per milligram protein at 37°C

(1983). This cDNA clone was used to screen a cDNA library from mRNA of human liver (Horwich et al. 1984). Takiguchi et al. (1984) isolated separately a cDNA clone, 1.6 kb long, for rat OTC. Horwich et al. (1984) and Takiguchi et al. (1984) determined the nucleotide sequences and deduced amino acid sequences of a mature subunit and its precursor of human and rat OTC respectively. There were 93% and 89% identities, in the amino acid and nucleotide sequences respectively, between rat and human OTC. The leader sequence, which consisted of 32 amino acids, contained no acidic amino acids, no long "membrane-spanning" hydrophobic regions, and five and eight basic amino acids in human and rat OTC respectively. Identity of the leader sequences between rat and human was 69% and 81% in the amino acid and nucleotide sequences respectively. These results suggest that not the leader primary sequence itself, but some other factors, such as the basic amino acid residues or the secondary structure, determine the mitochondrial distribution of the enzyme, since most other nuclear-encoded mitochondrial proteins have similar basic and nonhydrophobic precursor sequences. Horwich et al. (1985) actually demonstrated the importance of arginine residues in the leader sequence in experiments with in vitro mutagenesis.

Using an in situ hybridization technique, Lindgren et al. (1984) demonstrated that the gene for OTC is located on the short arm of the X chromosome, Xp21.1, which is near the Duchenne muscular dystrophy locus. This was further confirmed with somatic cell hybrid lines carrying various structural rearrangements of the human X chromosome short arm (de Martinville et al. 1985).

Hata et al. (1986) analyzed the structure of OTC gene by isolating phage clones carrying the OTC gene from human genomic DNA libraries and found that the OTC gene occupies a region of at least 85 kb (more than 120 kb, personal communication) on human genomic DNA.

4.2 Deficiency

Deficiency of OTC, the most common hereditary disorder of the urea cycle, causes hyperammonemia with increased urinary excretion of orotate. There are nonspecific increases in serum glutamine and alanine. Metabolic conversion of ammonia nitrogen into these amino acids in a patient with OTC deficiency was analyzed with ^{15}N by Kodama et al. (1983). The most characteristic feature of OTC deficiency is derived from its X-linked inheritance. Neonatal-onset lethal cases all occur in hemizygous males showing almost null OTC activity in the liver. Female patients show a wide variety of clinical manifestations, which is consistent with the Lyon hypothesis of random inactivation of the X chromosome.

4.2.1 Characteristics of the Enzyme Defect

Briand et al. (1982) classified OTC deficiency of hemizygous male patients into five groups of mutation. They analyzed liver OTC by determining enzymatic activity, kinetic properties, and enzyme amount with an immunological method (Table 4). Group 1 includes those who show undetectable activity accompanied with undetectable CRM in the liver. Group 2 comprises those who have no OTC activity but a significant amount of CRM in the liver. The patients in these two groups have symptoms of neonatal onset and would die soon after birth without treatment. Those in groups 3, 4, and 5 of late onset have partial activity. Group 3 is made up of those who have about 5% of OTC activity with normal kinetic properties, and the decreased OTC activity corresponds to the amount of CRM. Groups 4 and 5 comprise those who have a variant form of OTC which shows a larger Km value for ornithine and an abnormal curve of activity vs pH; the differences between the two groups are that variant OTCs of group 4 have a larger Km value for ornithine than those of group 5 and that the amount of CRM is decreased in group 4 but increased in group 5. Six patients belonged to group 1, followed by five in group 3 and three in group 4.

Table 4. Type of enzyme abnormalities in ornithine transcarbamylase (OTC) deficiency proposed by Briand et al. (1982)

Group	OTC activity	OTC protein (CRM)	Km value for ornithine	Optimal pH	Number of cases[a] A	B
1	0%	0%	–	–	6	4
2	0%	present	–	–	2	0
3	5%–30%	5%–30%	normal	normal	5	2
4	decreased	decreased	increased (× 10)	shift to alkaline	3	5
5	decreased	increased	increased (× 5)	shift to alkaline	1	0

[a] A and B are the number of cases given by Briand et al. (1982) and in our laboratory (Inoue et al., in preparation) respectively

In the authors' laboratory, however, of 11 male patients with OTC deficiency, five were considered to belong to group 4 and four to group 1 (Inoue et al., in preparation).

In our experiments with group 4, the mutant OTCs demonstrated very similar kinetic abnormalities (Yokoi et al. 1981; Inoue et al. in preparation), that is, about 10 times normal Km values for ornithine, and optimal pH displayed an alkaline shift. The Km values were similar but not identical. This suggests that each of the point mutations at different positions in the exon of the OTC gene makes a different amino acid substitution which, however, probably results in similar distortions of the OTC protein. The situation in group 5 may be similar even though groups 4 and 5 may differ in stability. This was further supported by the fact that sparse-fur mice, an animal model for OTC deficiency, have mutant OTC similar to group 5.

4.2.2 Analysis of mRNA and DNA

Some OTC deficiencies in group 1 have been analyzed at the mRNA level. Saheki et al. (1984) described a case of OTC deficiency in which OTC protein was undetectable by various immunological analyses, including immunoprecipitation followed by SDS-PAGE and immunobinding assay of CRM after electrophoretic transfer from polyacrylamide gel to nitrocellulose. In spite of the absence of CRM in the liver, the presence of mRNA coding for CRM with a molecular weight indistinguishable from the precursor of OTC was demonstrated by SDS-PAGE fluorography of newly synthesized peptides in an in vitro translation system with total RNA extracted from the liver of the patient. Furthermore, CRM of precursor size synthesized in a cell-free system was incorporated into rat liver and kidney mitochondria and processed to form a peptide indistinguishable in molecular weight from mature OTC. This suggests that the primary defect of OTC deficiency is located in the structural gene and that the labile OTC-related peptide, after being synthesized with its mRNA, was degraded too rapidly to be detected by the methods used.

Kodama et al. (1986) reported a male patient with no detectable CRM in the liver and where no translatable mRNA was detected with a cell-free translation system. They described further a female patient in whom a peptide with a smaller molecular weight was synthesized besides a faint normal peptide in vitro and was not transported into rat liver mitochondria.

Sparse-fur mice with abnormal skin and hair (spf-ash mice) are another animal model with deficient OTC which shows normal kinetic properties (Briand et al. 1982). This model corresponds to group 3 of human OTC deficiency; in other words, spf-ash mice have a decreased amount of OTC protein with molecular weight and kinetic properties comparable to those of controls. A cell-free translation system with RNA extracted from the liver of spf-ash mice (Briand et al. 1983; Rosenberg et al. 1983) revealed that the decreased amount of OTC protein was caused by a decreased level of translatable mRNA, and that two discrete in vitro products were found, one of normal precursor size and the other larger. Both products were taken up and processed to the mature form. The processed form of the larger product, however, did not bind to an affinity column containing δ-*N*-(phosphonacetyl)-*L*-ornithine, suggesting that it could not be converted to an active form and would be rapidly degraded.

Although it is difficult now to interpret the pathogenesis of these abnormal mRNAs at the molecular level, it may be assumed from the similarity to thalassemia (Forget et al. 1983; Orkin 1983) that the first case noted by Kodama et al. (1986) was caused by a transcriptional defect or a gene deletion, and that the second was due to a mutation in the exon making a premature termination codon or to a mutation which creates an alternative intron-exon splicing site, maybe as in spf-ash mice. In the case of heterozygous females, one cannot distinguish the between group 1 and group 3 enzyme defects.

Analysis of OTC deficiency at the genomic DNA level will be fruitful, as will be that of CPS deficiency, because one can diagnose them at any stage with any kinds of cells in which these two enzymes are not produced. Fetal liver biopsy may be possible, as reported by Rodeck et al. (1982) and Holzgreve and Golbus (1984). Nevertheless, it is virtually impossible to perform this procedure in humans because of its inherent danger.

Rozen et al. (1985) performed DNA analysis of 15 patients with OTC deficiency, using nearly full-length cloned human cDNA. Only one of them was found to have a partial deletion of the OTC gene at the 3′ portion. In addition, two distinct RFLPs were discovered at the OTC locus using the restriction endonuclease *MspI,* which made it possible in some OTC-deficient families to perform prenatal diagnosis and carrier detection.

A newly-formed *TaqI* restriction site in the OTC gene responsible for OTC deficiency was found in unrelated patients of different ethnic backgrounds (Nussbaum et al. 1986). This *TaqI* alteration yielded an aberrant 3.4 kb band on Southern blotting analysis which was not found in 84 control X chromosomes, ruling out the possibility of RFLP. This was used for prenatal diagnosis. Further intense analysis of genomic DNA will find new RFLPs, facilitating prenatal diagnosis and carrier detection.

5 Argininosuccinate Synthetase

5.1 Biochemistry

Argininosuccinate synthetase (ASS) is the least active enzyme in the urea cycle, catalyzing the formation of argininosuccinate from citrulline and aspartate with the released energy of ATP cleavage to AMP and pyrophosphate.

ASS has been purified from steers (Rochovansky et al. 1977), rats (Saheki et al. 1975), and human (O'Brien 1979; Kimball and Jacoby 1980; Saheki et al. 1983a). Rat and human enzymes cannot be distinguished on the basis of their antibodies with Ouchterlony double immunodiffusion, but there are slight differences in quantitative titration. The ASS consisted of four identical subunits with molecular weights of 43 000–48 000 (Saheki et al. 1977b, 1983a; O'Brien 1979; Rochovansky et al. 1977).

There are some discrepancies among reported papers concerning the kinetic properties of the enzymes (Table 5). Rochovansky et al. (1977), Kimball and Jacoby (1980), and Saheki et al. (1983a) observed nonlinear double reciprocal plots for the substrates of the steer, rat, and human enzymes, but Raushel and Seiglie (1983), and O'Brien (1979) found no such kinetics with the steer and human enzymes.

Takada et al. (1979) and Saheki et al. (1983a) found multiple forms of rat and human ASS under ion-exchange chromatography, PAGE, and isoelectric focusing. Multiple forms of rat ASS differed in the amount of bound argininosuccinate, although no changes in the elution profiles under ion-exchange chromatography occurred after the removal of bound argininosuccinate. On isoelectric focusing in the presence of urea as a denaturating agent, however, ASS migrated as a single band, suggesting that the multiplicity is due to the binding of some small molecules other than argininosuccinate.

Complementary DNA for ASS was cloned by Su et al. (1981) with RNA isolated from canavanine-resistant human cells which overproduce ASS protein, and with a differential filter hybridization technique. With the cDNA, the amino acid sequence for the human enzyme was deduced (Bock et al. 1983), the structure of the ASS gene was studied (Freytag et al. 1984a; Jinno et al. 1984, 1985), and abnormal RNAs from citrullinemic patients (Su et al. 1982, 1983; Kobayashi et al. 1986b, in preparation) were analyzed.

Southern blot analysis of genomic DNA with the cDNA probe revealed a large number (15–30) of hybridizing fragments with various restriction enzymes (Freytag et al. 1984a). The expressed gene for

Table 5. Physicochemical properties and kinetic parameters of purified human argininosuccinate synthetase (ASS)

	Liver ASS		Lymphoblast ASS
	O'Brien (1979)	Saheki et al. (1983a)	Kimball and Jacoby (1980)
Molecular weight			
subunit [a]	43 000	45 000	45 000
native enzyme	183 000	185 000	180 000
pI	–	7.6–8.2 (multiple)	–
Specific activity [b]	4.2 (37°C)	1.3 (25°C)	0.9 (37°C)
Km values (m*M*) for			
citrulline	0.016	0.024	0.061
aspartate	0.017	0.019	0.078
ATP	0.041	0.077 (0.26)[c]	0.094 (0.47)[c]

–, not described

[a] The subunit molecular weight deduced from cloned cDNA sequence was shown to be 46 400 (Bock et al. 1983)

[b] The specific activity, expressed as µmoles per minute per milligram protein, was determined with different methods and at different temperatures

[c] Two Km values for ATP were derived from the biphasic double-reciprocal plots

ASS is thought to be single (Carritt 1977; Beaudet et al. 1986); thus, most of the fragments are derived from 14 pseudogenes (Freytag et al. 1984a; Su et al. 1984), which were mapped onto 11 different chromosomes. Structural analysis of the pseudogenes revealed that they are dispersed processed pseudogenes of ASS (Freytag et al. 1984b). A recent report by Nomiyama et al. (1986) indicates that the ASS pseudogenes are not only derived from the functional gene via RNA intermediates, but also generated by gene duplication of some ancestor pseudogenes.

5.2 Deficiency: Citrullinemia

Deficiency of ASS causes citrullinemia, of which molecular defects have been extensively studied by Beaudet's group. Recently, Beaudet et al. (1986) published a review article under the title "The Human Argininosuccinate Synthetase Locus and Citrullinemia". In the present review, we make special reference to citrullinemia in Japan which has many characteristics not shared by the so-called classical citrullinemia first described by McMurray et al. (1962).

5.2.1 Three Types of Enzyme Abnormality in Citrullinemia

So far 51 cases of citrullinemia have been analyzed in our laboratory (Saheki et al. 1981, 1982a, 1982b, 1985a; Sase et al. 1985; Kobayashi et al. 1986b). Forty were adults aged from 18 to 66 years. Most suffered from sudden disturbance of consciousness such as disorientation, restlessness, and coma. Laboratory findings revealed no liver dysfunction, but hyperammonemia, especially at night, and also high concentrations of serum citrulline. Consanguinity was found in about 20% of them, suggesting an autosomal recessive form of inheritance.

Enzymatic and immunological analyses of the tissue specimens from the 51 citrullinemic patients revealed three different types of enzyme abnormality (Table 6) (Saheki et al. 1985a). Type I was qualitative; ASS displayed abnormal kinetic properties, such as higher Km values for the substrates and abnormal cooperative properties. Kinetically abnormal enzymes have been found in the liver, the kidney, and cultured skin fibroblasts and probably occur in all the cells containing ASS of patients with type I citrullinemia. The specific activity of the hepatic enzyme of one type I patient was calculated to be lower than that of controls from the data on the amount of ASS protein determined by enzyme immunoassay (Imamura et al. 1987). These results suggest that the enzymes in the various organs or cells originate from a single gene. The pathogenesis is presumably in the abnormal structure of the enzyme derived from a point mutation in the exon of the gene.

Type II was quantitative; the decrease of ASS activity in the liver was caused by a decrease in the amount of the kinetically normal enzyme. The most prominent characteristics of type II citrullinemia are that a decrease of ASS is observed in the liver but not in the kidney or cultured skin fibroblasts, in spite of the same genetic origin of the enzyme as that described above. Most adult citrullinemic patients (38 of 40) belonged to type II.

In type III citrullinemia no ASS activity was detected in the liver, the kidney, or cultured skin fibroblasts even when much higher concentrations of the substrates were used for the assay. A very small amount of the CRM, however, could be detected in the liver of type III by means of a sensitive enzyme immunoassay (Imamura et al. 1987). Four neonatal patients but none of the adult citrullinemic patients belonged to type III (Saheki et al. 1985a).

Table 6. Type of enzyme abnormalities in citrullinemia in Japan

	Type I	Type II	Type III
Frequency			
adult	1–2/40	38/40	0/40
infant	5/11	2/11	4/11
Consanguinity	2/6–7	8/38	2/4
Serum amino acid			
citrulline	very high	high	very high
arginine	low	high	low
Liver ASS			
activity	low	low	undetectable
enzyme amount	normal or low	low	very low
kinetics	abnormal	normal	–
mRNA coding for ASS			
content	n.d.	normal	variable
Northern blot	n.d.	normal	variable
activity	n.d.	normal or slightly low	undetectable or n.d.
distribution in the lobulus	n.d.	homogeneous or heterogeneous	negative
Kidney and fibroblast ASS			
activity	low	normal	undetectable
kinetics	abnormal	normal	–

ASS, argininosuccinate synthetase; n.d., not determined

5.2.2 Analysis of mRNA

In type II and III citrullinemia, the levels of ASS protein or CRM are reduced or very low, which means that analysis of mRNA is valuable for understanding their pathogenesis.

Dot blotting, Northern blotting, and S1 nuclease analyses revealed almost no reduction in ASS content and no gross structural abnormalities in the mRNA coding for ASS in the liver of type II citrullinemic patients (Kobayashi et al. 1986b). Also, experiments with a cell-free translation system showed almost normal translational activity of the mRNA for ASS (Sase et al. 1985). In these experiments, aldolase B, which is located in the cytosol of the liver and has a molecular weight similar to that of

ASS, was used as an internal control because constant recovery of mRNAs from human livers taken under various conditions is impossible.

Messenger RNA analysis of type III citrullinemic patients has so far yielded two kinds of results; reduced mRNA for ASS was found in the livers of two patients (Saheki et al. 1985a; Kobayashi et al. 1986b), and normal mRNA was found in the liver of another patient (Kobayashi et al., in preparation). No translational activity was found in an experiment with a cell-free translation system using the ASS mRNA of reduced content. We could not further investigate the abnormality of the reduced mRNA.

Further analyses with Northern blotting and S1 nuclease of the latter case showed a shorter mRNA with a defect of about 100 bases in the coding region near the 3′end (Fig. 2). This abnormal mRNA codes an ASS-like protein (CRM), which was detected in the liver of the patient by a sensitive enzyme immunoassay. The CRM, about 0.5% of the control content of ASS in the liver, was shown probably to hold a quaternary structure, since it demonstrated a molecular weight indistinguishable from the normal ASS on gel filtration chromatography. This suggests that the C-terminal region of ASS is necessary for the stability of the enzyme, but not for the formation of the quaternary structure.

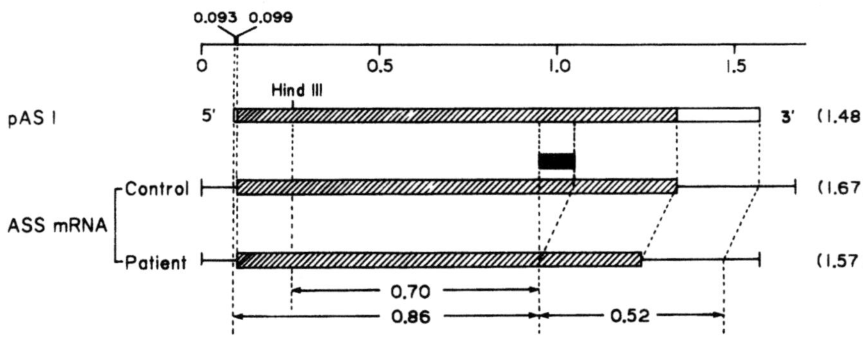

Fig. 2. Schematic representation of the supposed structure of the mRNA from a type III citrullinemic patient. This structure was deduced from the following results (Kobayashi et al., in preparation): argininosuccinate synthetase *ASS*-related mRNA about 1.57 kb long was detected in a poly$(A)^+$ RNA fraction of the liver with Northern blot analysis; two bands of 0.86 kb and 0.52 kb were found with S1 nuclease analysis using a cDNA clone, pAS-1; a 0.7 kb band was detected with direct S1 nuclease analysis using *Hind*III-cleaved and 3′-end-labeled pAS-1 according to the method described by Su et al. (1983)

5.2.3 Comparison with Cases Analyzed by Other Research Groups and Pathogenesis of Citrullinemia

The abnormalities of ASS protein in type I citrullinemia are qualitative ones, as described by Kennaway et al. (1975) and Lockridge et al. (1977), and are probably caused by a point mutation in exon of the gene.

Type III citrullinemia may correspond to or be included in a group of neonatal citrullinemia analyzed by Beaudet et al. (1986) with Northern blotting and S1 nuclease mapping. They carried out analyses of immunoreactive protein, RNA, and DNA in a series of cultured skin fibroblasts from neonatal citrullinemic patients. All of these cell lines had less than 1% of ASS activity, and the majority demonstrated the absence of CRM on Western blotting analysis. Beaudet et al. (1986) eliminated the possibility of deletion of the entire gene in their patient analyzed with Southern blotting of the genomic DNA. Analyses of RNA from these cultured fibroblasts with Northern blotting and S1 nuclease mapping indicated multiple classes of abnormal mRNA, gross structural abnormalities found by Northern blotting and S1 nuclease digestion, mRNA with no detectable abnormality probably containing single-base changes, and an absence of mRNA.

The shorter mRNA with a deletion of about 100 bases analyzed in our laboratory may be derived from a partial deletion of the genomic DNA or, more probably, from one base substitution resulting in dysfunction of RNA splicing.

The pathogenesis of Type II citrullinemia is more difficult to understand. In type II citrullinemia, decreased hepatic ASS of normal kinetic properties and heat stability comparable with the controls (Akaboshi et al. 1983) is accompanied with normal amount of mRNA coding for ASS in the liver, no gross structural abnormality so far determined, and translational activity comparable with the controls. The mRNA may contain some point mutation in the coding region which does not affect kinetic properties but makes the enzyme unstable in the liver. But this hypothesis must explain the reason why the kidney and cultured skin fibroblasts contain normal levels of ASS. Alternatively, some mutation may make the mRNA translate in vivo slower than at the normal rate.

At least three reports apart from ours have been published so far on cases of citrullinemia very similar to our type II citrullinemia; one from Japan showed about half the normal activity of ASS in the liver with normal activity in the leukocytes (Kooka et al. 1977), and two from Europe (Vidailhet et al. 1971; Roerdink et al. 1973), showed no activity in the liver with normal activity in the kidney and brain. These reports suggest that type II citrullinemia is not limited to Japan.

5.2.4 Laboratory Findings in Type II Citrullinemia

It has been reported that hyperammonemia is not present consistently in citrullinemia and, if present at all, occurs only during the postprandial period (Shih 1976). In type II citrullinemic patients, blood ammonia frequently shows diurnal rhythms, probably due to dietary intake; it is low in the morning and increases in the evening, and the diurnal rhythms are abolished during fasting (Yajima et al. 1982).

Serum citrulline levels of type II citrullinemic patients (5–30 times those of the controls) are not as high as those of type I and III patients (usually more than 40 times those of controls), and often fluctuate during their clinical courses (Saheki et al. 1986).

One difference between type II and the other types is that, as shown in Fig. 3, the serum arginine levels of type II citrullinemic patients are higher than in controls, whereas type I and III patients are arginine-deficient (Saheki et al. 1985b, 1986).

This phenomenon can be explained by the difference of ASS activity in the kidney between type II and the other types, and by a widely accepted theory (Windmueller and Spaeth 1981; Funahashi et al. 1981) about the origin of the circulating arginine. According to the theory, circulating arginine is derived from arginine synthesized by the action of ASS and argininosuccinate lyase (ASL) in the kidney from citrulline, which is synthesized from ornithine in the intestine under normal conditions. Under citrullinemic conditions, citrulline is released in large quantities from the liver by the defective action of the urea cycle.

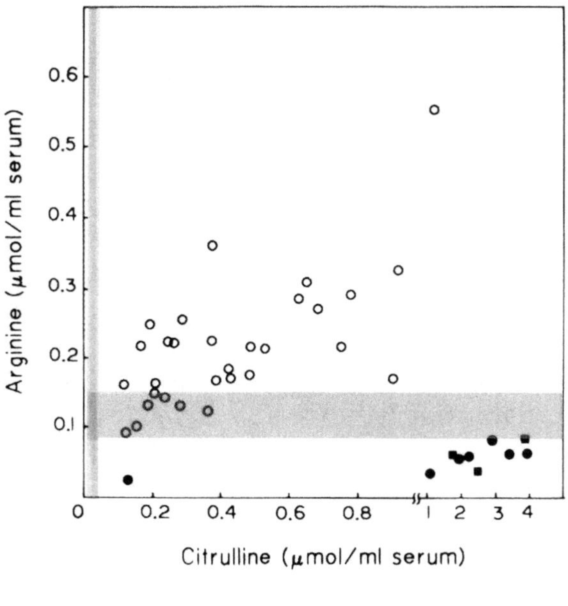

Fig. 3. Relation between serum citrulline and arginine of type I (●), type II (○), and type III (■) citrullinemic patients. ***Shaded areas*** indicate control ranges of serum citrulline and arginine

Citrulline accumulated in circulation cannot be converted to arginine in the case of type I and III patients because they have a defect of ASS in the kidney, but citrulline is converted efficiently to circulating arginine in the kidney of type II patients because they have normal renal ASS activity.

It is also a prominent characteristic of type II citrullinemia that the serum levels of alanine, serine, glycine and branched-chain amino acids are significantly lower than in the controls (Saheki et al. 1986), in spite of the fact that serum alanine levels are usually higher in hyperammonemia (Shih 1976). As a result, the ratio of threonine and serine is characteristically higher, and that of branched-chain amino acids to aromatic amino acids lower, than the controls (Saheki et al. 1986).

A characteristic change of citrullinemia in urinary amino acids apart from citrulline is an increased excretion of homocitrulline (Shih 1976). A remarkable increase in homocitrulline excretion was also observed in type II citrullinemia (406±217 nmol/mg creatinine, n=11, as compared to the control value of 28.5±10.5 nmol/mg creatinine, n=9; Saheki et al., unpublished data). The mechanisms underlying such changes in serum and urinary amino acids apart from citrulline and arginine remain unclear, although disturbances in lysine metabolism and the presence of an alternative urea cycle are postulated for the increased excretion of homocitrulline in citrullinemia (Levin et al. 1973; Scott-Emuakpor et al. 1972; Matsuda et al. 1979).

5.2.5 Localization of Argininosuccinate Synthetase in the Lobulus and its Abnormality in Type II Citrullinemia

The urea cycle enzymes are located at the periportal region of the lobulus (Mizutani 1968; Gaasbeek Janzen et al. 1981; Gebhardt and Mecke 1983; Häussinger 1983). In our experiments with rat liver using immunohistochemical methods (Kobayashi et al., in preparation), ASS and arginase were present in the cytoplasm of the hepatocytes almost throughout the lobulus except for two or three hepatocyte layers around the hepatic vein. This is in contrast to the reported localization of glutamine synthetase (Gebhardt and Mecke 1983). This clear patchwork distribution of the urea cycle enzymes and glutamine synthetase may play a role in the nitrogen metabolism of the liver (Häussinger 1983).

One question was whether the decreased ASS content in the liver of type II citrullinemic patients represented an even decrease in all hepatocytes or an uneven decrease.

The answer was as follows (Saheki et al. 1983b). There were mainly two types of ASS protein distribution in the liver of type II citrullinemic patients. In some cases, most of the hepatocytes were stained homogeneously with the antisera to ASS, as in the control liver (homogeneous type); in about half of the type II liver specimens, however, ASS was distributed unevenly among the hepatocytes: there was a clustered type (Fig. 4), in which the positively stained hepatocytes formed clusters among the poorly stained cells, and, more rarely, a disseminated type, in which the positively stained hepatocytes were distributed among the poorly stained cells. We consider that the disseminated distribution may be a transient form of the clustered distribution.

The clustered type of distribution of ASS was seen only in type II citrullinemia, not in other types of citrullinemia, hepatitis, liver cirrhosis, or hepatoma (Yagi et al., in preparation). Furthermore, the clustered enzyme distribution was observed only with ASS, not with arginase, aldolase B, and γ-glutamyltranspeptidase in type II citrullinemic livers.

a

b

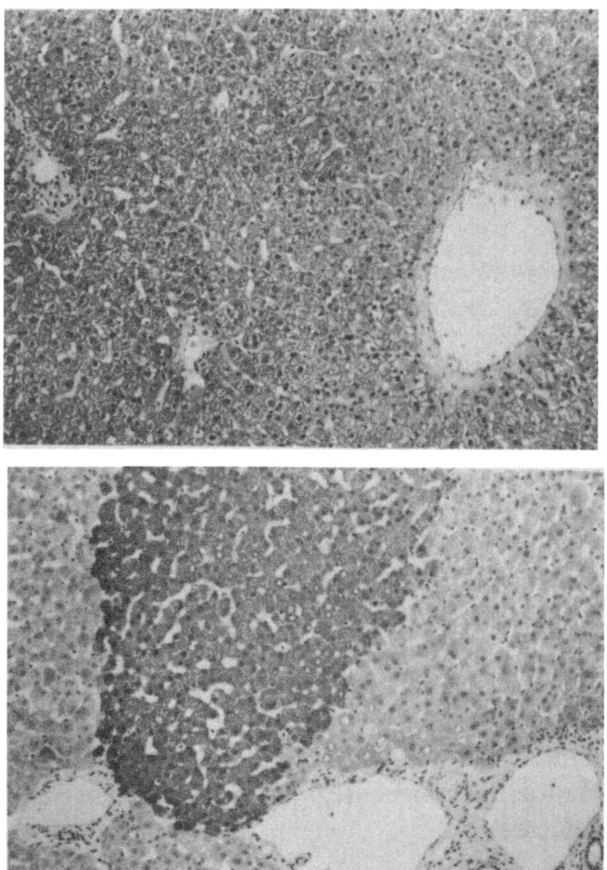

Fig. 4a, b. Immunohistochemistry of argininosuccinate synthetase in the liver of a control subject (*a*), showing homogeneous distribution, and a type II citrullinemic patient (*b*), showing clustered distribution. The sections were counterstained with hematoxylin after the immunostaining. No relation was observed between the cluster formation and liver lobulus

Among 18 cases of type II citrullinemia that were examined, 11 showed a clustered distribution of ASS in the lobulus. There were no differences between the homogeneous and clustered types in age, sex, hepatic ASS activity, and serum citrulline levels. Nevertheless, the clinical prognosis was quite different; nine of 11 cases belonging to the clustered type died, mostly of brain edema, but all seven patients belonging to the homogenous type are still alive.

The cause of the uneven distribution of ASS in the liver of type II citrullinemic patients and its relation to the pathogenesis of type II citrullinemia remain unresolved.

5.3 Prenatal Diagnosis of Citrullinemia

There are many difficulties in prenatal diagnosis of urea cycle enzyme deficiencies: CPS and OTC are expressed almost exclusively in the liver, and the other urea cycle enzymes, too, are expressed mainly in the liver not in early, but in late fetal life. Although there have been a few reports on prenatal diagnosis with fetal liver biopsy (Rodeck et al. 1982; Holzgreve and Golbus 1984), this form of diagnosis does not seem to have caught on.

Christensen et al. (1980) and Jacoby et al. (1981) reported that the ASS activity in the amniotic cells used for the prenatal diagnosis of citrullinemia was low and variable – very low in the epithelial-like cells, higher in the fibroblast-like cells – while ASL activity was invariable regardless of cell type. Cathelineau et al. (1981a) circumvented this problem by measuring incorporation of ^{14}C-citrulline and ^{3}H-phenylalanine into cell protein fraction and comparing the $^{14}C/^{3}H$ ratios in 6 h and 24 h incubations. It has been reported (Fleisher et al. 1983; Kamoun et al. 1983; Kleijer et al. 1984) that a high concentration of citrulline in amniotic fluid has considerable diagnostic value. Christensen et al. (1985) suggested that assay of ASS in chorionic villi taken during the first trimester of pregnancy would be useful in citrullinemia diagnosis. All these methods may be useful for the diagnosis of type I and III citrullinemia, but not type II.

A possible future tool for prenatal diagnosis of citrullinemia is DNA analysis. But the difficulty here was shown by Su et al. (1984): because of the presence of numerous pseudogenes, many DNA fragments were found on Southern blot analysis of human genomic DNAs when an ASS cDNA was used as a probe.

Jinno et al. (1986) found RFLPs at the ASS gene locus. They used as probes three DNA fragments prepared from the 5′-end region, not the exon, of the ASS gene, which do not share common DNA sequences with ASS pseudogenes. Southern blot analysis of *AluI* fragments of genomic DNA with the ASS gene-specific probes demonstrated two alleles in Japan. Jinno et al. also suggested an RFLP produced by *HinfI*. These RFLPs will help prenatal and carrier detection, especially of type II patients, who show abnormal ASS activity only in the liver.

6 Argininosuccinate Lyase

6.1 Biochemistry

Argininosuccinate lyase (ASL) is a cytosolic urea cycle enzyme successive to ASS, catalyzing the cleavage of argininosuccinate into arginine and fumarate. Although all the human tissues examined exhibit significant ASL activities (Ratner 1973), the liver and kidney are the major organs containing ASL, which carries out urea synthesis and arginine formation from citrulline respectively.

ASL has been purified from bovine liver (Havir et al. 1965), kidney (Bray and Ratner 1971), and brain (Murakami-Murofushi and Ratner 1979) and human liver (O'Brien and Barr 1981; Palekar and Mantagos 1981). The physicochemical and catalytic properties of the bovine enzymes have been extensively studied and reviewed by the Ratner group (Ratner 1973, 1976). Those of purified human ASL are given in Table 7.

Identity of ASL from various organs of steer was strongly suggested by analyses of protein-chemical, immunochemical, and enzymological properties published by Murakami-Murofushi and Ratner (1979). Similar immunochemical analysis was performed on human enzymes from the liver and fibroblasts by O'Brien and Barr (1981). Also, Kobayashi et al. (1986a) showed that antisera raised against human liver ASL could not differentiate ASL proteins in the liver, kidney, brain, and erythrocytes on immunotitration tests. On the other hand, the antigenicity of ASL proteins from different species (human, rat, and steer) is considerably different (O'Brien and Barr 1981; Kobayashi et al. 1986a), while CPS and ASS show very similar antigenicity between species (see Sect. 3 and 5).

Recent advances in ASL analysis are the cloning of cDNAs for rat (Kawamoto et al. 1985; Amaya et al. 1986; Lambert et al. 1985, 1986) and human (Kalumuck et al. 1985) enzymes. Figure 5 shows a restriction map of rat ASL cDNA clarified by Amaya et al. (1986). The human ASL gene was mapped onto chromosome 7 with the human cDNA in human x Chinese hamster somatic cell hybrids (Kalumuck et al. 1985), which

Table 7. Physicochemical properties and kinetic parameters of purified human argininosuccinate lyase

Parameters	O'Brien and Barr (1981)	Palekar and Mantagos (1981)
Molecular weight		
subunit	49 000	50 000
native enzyme	187 000	200 000
Optimal pH		
forward reaction	7.5	–
reverse reaction	6.8–7.2	–
Specific activity [a]		
forward reaction	10.3	20.0
reverse reaction	8.0	–
Km values (m*M*) for		
argininosuccinate	0.20	0.1
fumarate	5.3	–
arginine	3.0	–

–, not described
[a] μmoles per minute per milligram protein at 37°C or 38°C

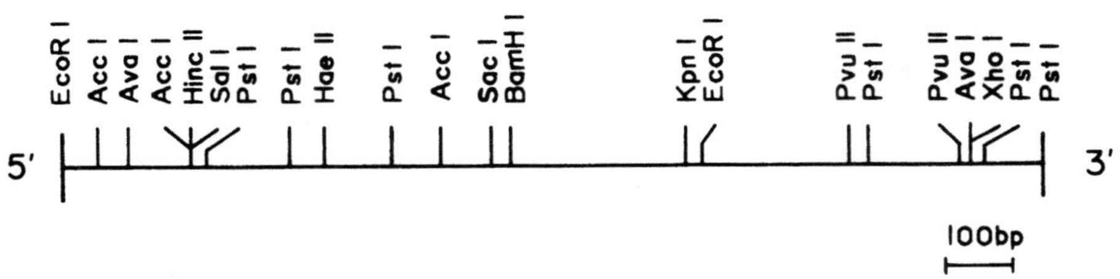

Fig. 5. Restriction map of rat liver argininosuccinate lyase cDNA

confirmed the earlier result obtained by bioautography with gel electrophoresis and auxotrophic bacteria (Naylor et al. 1978). No RFLPs were found in Southern blots of human genomic DNA cut with 18 different restriction enzymes and proved with cDNA (Kalumuck et al. 1985).

6.2 Deficiency: Argininosuccinic Aciduria

Argininosuccinic aciduria, first described by Allan et al. (1958), is the second most commonly reported disorder of the urea cycle and is characterized by accumulation of argininosuccinate or its anhydride in blood, urine, and cerebrospinal fluid. Clinically, it is divided into three types (Shih 1978); in the early-onset neonatal type, symptoms such as respiratory distress, poor feeding, seizure, and lethargy begin within the first few days of life; in the subacute and late-onset types, the major symptom is mental retardation in addition to vomiting, seizure, and hepatomegaly. Hair abnormality, trichorrhexis nodosa, is another characteristic of argininosuccinic aciduria.

In Japan, ASL deficiency is not as frequent as in the United States, where the reported incidence is 1 in 70 000 live births (see Walser 1983); six cases have been reported in Japan so far (Sakiyama et al. 1982; Yamamoto et al. 1983; Oyanagi et al. 1985; Kamimura et al. 1986; Kobayashi et al. 1986a; Akaboshi et al., submitted).

The most remarkable aspect of ASL deficiency in argininosuccinic aciduria is the variation of the enzyme defect among organs and cells. As listed in Table 8, several reports described patients with no detectable hepatic ASL activity, but no decrease in, or at least significant presence of ASL activity in the kidney or brain (Glick et al. 1976; Perry et al. 1980). This enzymatic variation may be supported by laboratory findings showing that serum arginine levels are not low, but higher in some argininosuccinic aciduria patients (Quereshi et al. 1978). The situation seems to be very similar to that of type II citrullinemia described in Sect. 5 above. So far, however, there have been no reports of the enzymological properties or protein nature of the deficient ASL. In a few cases, no organ or cell examined had ASL activity (Kobayashi et al. 1986a). These results, together with complementation analysis (Cathelineau et al. 1981b; McInnes et al. 1984) and the established localization of ASL structure gene on chromosome 7 (Naylor et al. 1978; Lambert et al. 1986), suggest that some regulator gene might be present for the organ-specific expression of ASL protein, as supposed by Glick et al. (1976).

So far, only one report has described the level of deficient ASL protein in the liver of two argininosuccinic aciduria patients who had no detectable ASL activity (Kobayashi et al. 1986a). A complete or almost complete absence of ASL protein was revealed by immunotitration and Ouchterlony double immunodiffusion.

Further research on the mRNA and genomic DNA with cDNA will help us understand not only the pathogenesis of the disorder, but also the regulatory mechanisms for organ-specific gene expression.

Table 8. Argininosuccinate lyase activity in patients with argininosuccinic aciduria (cases in which argininosuccinate lyase activity in multiple organs or cells was determined)

Author(s)	Argininosuccinate lyase activity (% of controls)				
	Liver	Kidney	Brain	Fibroblast	Erythrocyte
Carton et al. (1969)	–	<0.6	2.7	–	–
Pollitt (1973)	7	–	–	<2	1
van der Heiden et al. (1976)	0.5	0	–	3.8	–
Farriaux et al. (1976)	<4	<7	<14	–	–
Glick et al. (1976)	0	135	195	–	16
Schutgens et al. (1979)	0	–	–	0	2.6
Perry et al. (1980)	0	4.2	49	14	–
Kobayashi et al. (1986a)					
Case 1	0	–	–	0	0
Case 2	0	0	0	–	0

–, not described

7 Arginase

7.1 Biochemistry

Arginase is the last enzyme of the urea cycle, forming urea, a final product of the cycle, and ornithine, a nitrogen carrier, for further cycling by the hydrolysis of arginine. Arginase activity is found not only in the liver, but also in the kidney, brain, intestine, salivary glands, mammary glands, erythrocytes, fibroblasts, and so on (Herzfeld and Raper 1976). The role of extrahepatic arginase is still unclear, although the enzyme is postulated to supply ornithine for the synthesis of polyamines and proline (Mezl and Knox 1977; Pohjanpelto and Holtta 1983; Carvajal and Cederbaum 1986) and to inhibit the growth of arginine-dependent cells such as lymphoblastoid cell lines and lymphocytes (Chen and Broome 1980).

Liver arginases have been purified from horse (Greenberg et al. 1956), rat (Schimke 1964; Hirsch-Kolb and Greenberg 1968), rabbit (Vielle-Breitburd and Orth 1972), and human (Bascur et al. 1966; Berüter et al. 1978; Brusdeilins et al. 1985). The molecular weight of human liver arginase is estimated to be 98 000–120 000 of probably a trimer structure with a subunit of 35 000–39 000 (Brusdeilins et al. 1985). The physicochemical and kinetic parameters of purified human liver arginase are listed in Table 9.

Multiple forms of arginase were found in various tissue preparations by ion-exchange chromatography and genetically different isozyme forms were confirmed by immunochemical analysis and analysis of argininemic patients (Kaysen and Strecker 1973; Herzfeld and Raper 1976; van Elsen and Leroy 1975, 1977; Michels and Beaudet 1978; Brusdeilins et al. 1985; Konarska et al. 1981, 1983, 1985; Spector et al. 1983). Kidney arginase, immunologically different from the liver enzyme,

Table 9. Physicochemical properties and kinetic parameters of purified human liver arginase

Parameters	Bascur et al. (1966) Carvajal et al. (1971)	Berüter et al. (1978)	Brusdeilins et al. (1985)
Molecular weight			
subunit	30 000	35 000	39 200±400
native	118 000	107 000	100 000
pI	9.2	–	8.9
Optimal pH	9.4–9.5	9.3	–
Specific activity[a]	116 (pH 9.0)	2090 (pH 9.5)	250 (pH 9.5)
Km value (m*M*) for			
arginine	6.2	10.5	–

[a] μmoles per minute per milligram protein at 37°C

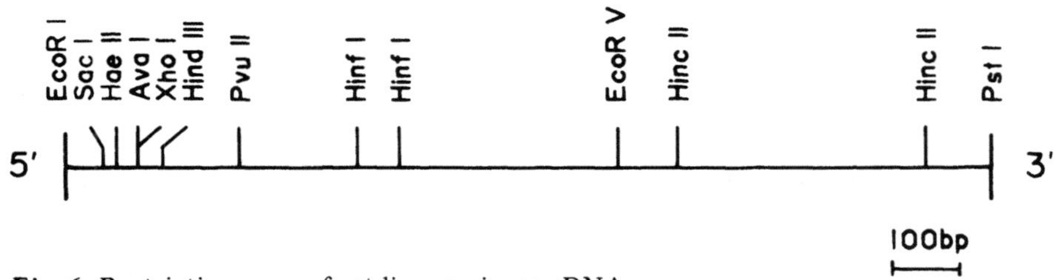

Fig. 6. Restriction map of rat liver arginase cDNA

has been purified only partially (Kaysen and Strecker 1973; Skrzypek-Osiecka et al. 1983).

Dizikes and Cederbaum (1984) isolated cDNA clones for rat liver arginase, with which a single mRNA species of about 1500 bases was shown to hybridize on Northern blot analysis of poly $(A)^+$ RNA extracted from rat liver. No detectable hybridization was found with mRNA from rat kidney and brain, whereas about 1% activity was derived from liver-type arginase. Similar results were obtained separately by Kawamoto et al. (1985, 1986). The isolated cDNA clone from a cDNA expression library constructed in a plasmid, pUC8, from poly $(A)^+$ RNA of rat liver contained an insert approximately 1.35 kb in length, of which a restriction map is shown in Fig. 6. A specific protein band of about 43 000 daltons was detected in SDS-PAGE of the sonicated extracts of the bacteria transformed with this cloning vector, pUC8, the product was presumably synthesized as a fused protein containing six amino acid residues derived from β-galactosidase and peptide from arginase (35 000–40 000). Thus, the product seems to contain all or most parts of the mature arginase, and indeed the transformed bacteria showed a high activity of arginase.

Dizikes et al. (1985) isolated cDNA clones for human liver arginase using rat liver cDNA clones. They described that Southern blot analysis of human genomic DNA revealed single bands of about 20 kb with several restriction endonucleases and up to five smaller bands with some other enzymes. From the results, Dizikes et al. (1985) suggested that there may be a single gene locus for liver arginase in man, possibly as little as 14 kb in length.

7.2 Deficiency: Argininemia

Argininemia caused by arginase deficiency is inherited in an autosomal recessive manner and is the least frequently diagnosed urea cycle enzyme defect, apart from *N*-acetylglutamate synthetase deficiency (Walser

1983). Since the two patients within a single family described by Terheggen et al. (1969), argininemia has been reported so far in a total of 20 patients in 15 families (summarized by Konarska and Tomaszewski 1986; Bernar et al. 1986; Kubo, personal communication). The small number of cases may result either from the low prevalence of the disease or from the absence of simple and reliable screening tests. The symptoms of argininemia are not very specific, do not begin at neonatal periods, and progress slowly, but the clinical pictures of argininemia reported so far have been remarkably uniform and unlike those seen in any other urea cycle enzymopathy (Walser 1983): irritability and marginally delayed development in early infancy; muscular hypertonicity of the lower extremities which usually progresses to spastic diplegia; vomiting, tremor, ataxia, and choreoathetosis which progresses finally to seizures, psychosis, and mental retardation.

7.2.1 Pathophysiology

The diagnosis of argininemia can be made on the basis of a high concentration of serum arginine, loading tests, and arginase deficiency in erythrocytes. A high concentration of arginine is observed in serum, urine, and cerebrospinal fluid, and only slightly in cord blood (Terheggen et al. 1975).

The urinary amino acid pattern of argininemia resembles that of cystinuria-lysinuria (Terheggen et al. 1969, 1975; Qureshi et al. 1981); large amounts of cystine, lysine, and ornithine are excreted in the urine by competitive inhibition of dibasic amino acids at the renal reabsorption mechanisms.

A variety of guanidino compounds, such as *N*-acetylarginine, α-keto δ-guanidinovaleric acid, argininic acid, and homoarginine, are found in quantities in the serum, urine, and cerebrospinal fluid (Marescau et al. 1985). Since some of them are experimentally epileptogenic (Marescau et al. 1985), the main symptoms may be caused by the accumulation of these guanidino compounds.

Another characteristic finding is an increased excretion of orotic acid, although hyperammonemia is rarely seen (Snyderman et al. 1977; Walser 1983). Orotic aciduria is considered to result from allosteric activation of *N*-acetylglutamate synthetase by arginine, which leads to stimulated synthesis of carbamylphosphate (Bachmann and Colombo 1980), and/or from decreased availability of ornithine for mitochondrial citrulline formation. The latter condition is caused by defective formation of ornithine from arginine (Qureshi et al. 1981; Yoshino et al. 1982) or by inhibition of mitochondrial ornithine uptake by arginine (Kang et al. 1983; Hommes et al. 1983b). This second pathophysiological consideration may be countered by ornithine supplementation therapy which causes normalization of orotic acid excretion (Kang et al. 1983).

7.2.2 Enzymatic and Immunochemical Analyses

In affected patients, deficiency of the enzyme activity is demonstrable in the liver (Michels and Beaudet 1978; Cederbaum et al. 1979b); erythrocytes (Terheggen et al. 1969; Cederbaum et al. 1979b; Snyderman et al. 1979; Michels and Beaudet 1978; Qureshi et al. 1983; Endres et al. 1984; Bernar et al. 1986; Konarska and Tomaszewski 1986), leukocytes (Cederbaum et al. 1979b; Marescau et al. 1979; Endres et al. 1984), stratum corneum (Michels and Beaudet 1978), and saliva (Konarska et al. 1985), but not in fibroblasts (van Elsen and Leroy 1977; Konarska et al. 1981; 1983) or the kidney (Spector et al. 1983). Indeed, it has been reported that kidney specimens obtained from argininemic patients had arginase activity 4–5 times greater than that found in normal kidney material (Spector et al. 1983).

There have been few reports on immunochemical analysis of argininemia. Spector et al. (1983) showed that antiserum against human liver arginase precipitated all of the enzyme activity in the liver and erythrocytes, but only part of that in the kidney, brain, and intestine. Furthermore, Spector et al. (1983) revealed the presence of enzymatically inactive CRM in the erythrocytes and kidney of argininemic patients by immunotitration analysis. These analyses of argininemia confirm that there are two gene loci determining arginase in man. It should be noticed that antiserum against rat liver arginase was not sufficient to recognize, at least, mutant human arginase protein as compared to anti-human serum (Spector et al. 1983; Bernar et al. 1986).

There may be specific situations in argininemia similar to those in citrullinemia and argininosuccinic aciduria mentioned above, where arginase is deficient in the liver but not in fibroblasts (van Elsen and Leroy 1977; Konarska et al. 1983); however, further extensive research must be done.

8 *N*-Acetylglutamate Synthetase

N-Acetylglutamate synthetase catalyzes the formation of *N*-acetylglutamate from acetyl-CoA and glutamate. The enzyme was discovered by Shigesada and Tatibana (1971a, 1971b) and was purified from human liver (Bachmann et al. 1982) and rat liver (Sonoda and Tatibana 1983). Considering the very low content of the enzyme in the liver and the

Table 10. Physicochemical properties and kinetic parameters of purified human and rat liver *N*-acetylglutamate synthetase

	Human enzyme (Bachmann et al. 1982)	Rat enzyme (Sonoda and Tatibana 1983)
Molecular weight		
subunit	–	57 000
native enzyme	190 000	160 000
Optimal pH	8.5	8.2
Specific activity [a]	2.86	18.7
Km values (m*M*) for		
glutamate	8.1	3.7
acetyl-CoA	4.4	0.76
Activator	arginine	arginine argininic acid protamine polyarginine polylysine

[a] Specific activity, expressed as μmoles per minute per milligram protein, was determined at 37°C with 4 m*M* glutamate and 2 m*M* acetyl-CoA for the human enzyme and at 25°C with 1 m*M* glutamate and 0.5 m*M* acetyl-CoA for the rat enzyme

specific activity of the enzyme purified from rat liver (Sonoda and Tatibana 1983), the purified human enzyme does not appear to be homogeneous. The physicochemical and kinetic parameters of the two purified enzymes are given in Table 10.

N-Acetylglutamate is a natural activator for CPS (Shigesada and Tatibana 1971b) and plays an important role in the short-term regulation of urea synthesis. The hepatic content of *N*-acetylglutamate fluctuates considerably under various physiological and experimental conditions (McGivan et al. 1976; Saheki et al. 1977a; 1978; Shigesada et al. 1978; Stewart and Walser 1980). The fluctuations of *N*-acetylglutamate are considered to result from the fluctuations of glutamate and arginine concentrations in the liver; the former is one of the substrates and the latter is an allosteric activator of the enzyme (Shigesada and Tatibana 1971a; 1978). Kawamoto et al. (1982, submitted) reported that the extent of the activation of the enzyme by arginine rises during postprandial periods and suggested that this is caused by oxidative modification of the enzyme sulfhydryl group.

So far, only one case of *N*-acetylglutamate synthetase deficiency has been reported, by Bachmann et al. (1981). According to their descriptions, the clinical symptoms and the laboratory findings were indistinguishable from those of CPS deficiency. In the biopsy liver specimens, the activity of CPS was found to be normal and that of *N*-acetylglutamate synthetase was undetectable.

They treated the patient with carbamylglutamate, an analogue of *N*-acetylglutamate. Reduction of the dose led to hyperammonemia. The fact that a case with *N*-acetylglutamate synthetase was found and that the deficiency caused hyperammonemia shows the importance both of *N*-acetylglutamate and of synthetase in urea synthesis.

9 Hyperornithinemia, Hyperammonemia, and Homocitrullinuria

9.1 Symptoms and Laboratory Findings

Since the first case reported by Shih et al. (1969), 11 cases of hyperornithinemia, hyperammonemia, and homocitrullinuria (HHH) have been published (Wright and Pollitt 1973; Gatfield et al. 1975; Hommes et al. 1982; Oyanagi et al. 1983; Yoshida et al. 1984; Simell et al. 1985). HHH patients, aged from 16 months to 41 years old, suffer from vomiting, seizures, ataxia, and mental retardation. They dislike a protein diet, in common with patients with other urea cycle enzymopathies.

Laboratory findings of HHH are, as shown by the name, increased serum concentration of ornithine with hyperammonemia accompanied by increased excretion of homocitrulline and orotic acid.

9.2 Pathophysiology

Hyperornithinemia, hyperammonemia, and homocitrullinuria can be explained largely by a deficient transport of ornithine into mitochondria. Cytoplasmically-generated ornithine has to be transported to mitochondria, where a degradative enzyme of ornithine, ornithine transaminase, and a urea cycle enzyme converting ornithine and carbamylphosphate to citrulline, ornithine transcarbamylase, are located. The suggested defect of mitochondrial ornithine transport in HHH may result in the accumulation of ornithine in cytoplasm and extracellular fluids owing to lack of ornithine degradation, and in hyperammonemia and increased excretion of orotic acid due to inoperation of the urea cycle and accumulation of carbamylphosphate.

Increased excretion of homocitrulline may result from stimulated synthesis of homocitrulline from lysine and carbamylphosphate, as suggested by the result of loading tests (Gatfield et al. 1975). OTC is able to catalyze the formation of homocitrulline from carbamylphosphate and lysine instead of ornithine. Fell et al. (1974) suggested that OTC can use lysine owing to a higher ratio of lysine to ornithine in the mitochondria of HHH patients, in spite of the very high Km value of OTC for lysine. Hommes et al. (1983b) reported that ornithine and lysine were transported into mitochondria by a common transport system and that a mutated transport protein in HHH was assumed to be still operative for lysine. Oyanagi et al. (1983), however, suggested the presence of an independent mitochondrial transport system for lysine from the analysis of a patient with hyperlysinemia, and Hommes et al. (1983a) and Carter et al. (1984) reported that lysine-specific transcarbamylase is present at the intermembrane space of mitochondria. Homocitrullinuria has been found not only in HHH, but also in citrullinemia (Buist et al. 1974; Shih 1976), saccharopinuria (Carson et al. 1968; Cederbaum et al. 1979a), hyperlysinemia (Armstrong and Robinow 1967), and lysinuric protein intolerance. Much further research needs to be done on homocitrullinuria.

9.3 Pathogenesis

Deficiencies of CPS (Gatfield et al. 1975) and ornithine decarboxylase (Shih and Mandell 1974) have been suggested as causative factors for HHH. But these enzyme deficiencies could not explain the pathophysiology. A defect of mitochondrial ornithine transport as a possible cause of HHH was first suggested by Fell et al. (1974) on the results of amino acids and protein loading tests. This has been supported by the experiments with the fibroblasts from HHH patients, in which the incorporation of labeled ornithine into cellular proteins as proline and glutamate residues (Shih et al. 1982; Gray et al. 1982, 1983; Hommes et al. 1982), and the conversion of labeled ornithine into proline, glutamate, aspartate, and CO_2 (Gray et al. 1982) were less than in the controls. These results indicate that the defect is expressed not only in the liver, but also in the fibroblasts.

Gray et al. (1983) suggested presence of a mutant protein in the fibroblasts which had a larger Km value for ornithine but Vmax comparable to the controls. The unusual structures of mitochondria seen in the hepatocytes and fibroblasts from HHH patients may result from a rearrangement of the mitochondrial membrane induced by the incorporation of such mutated protein into the inner membrane (Haust et al. 1981; Metoki et al. 1984).

There has been only one experiment done with the liver of an HHH patient which suggested a defect of mitochondrial ornithine transport; Oyanagi et al. (1983) revealed a decreased content of ornithine, expressed as ratios of ornithine to each of the other amino acids, in the isolated mitochondria.

We have recently had opportunities to analyze three cases of HHH enzymologically and metabolically with liver biopsy specimens (Inoue et al., in preparation). HHH patients no. 1, 2, and 3 were 36, 11, and 19 years old respectively. Clinical manifestations of all the patients were as stated above, and their serum and urinary concentrations of ammonia, amino acids, and orotic acid were indicative of HHH. Patient no. 3 had a younger brother who was also diagnosed as having HHH.

Activity of citrulline formation with ammonia, bicarbonate, succinate, and ornithine (5 m*M*) of the respirating mitochondria prepared from the liver of HHH patients was 20%–40% of that in controls, although no reduction was found in the activity of the hepatic urea cycle enzymes and in the concentration of *N*-acetylglutamate was found. All the mitochondria from the HHH patients showed larger Km values for ornithine but variable Vmax values – lower, normal, and higher values were obtained (one of them is presented in Fig. 7), indicating the presence of a variety of mutant transport proteins.

Direct measurements of the initial rate of ornithine transport by centrifugal filtration described by McGivan et al. (1977) and Hommes et al. (1982) with ^{14}C-ornithine and ^{3}H-sucrose for 5–20 s incubation

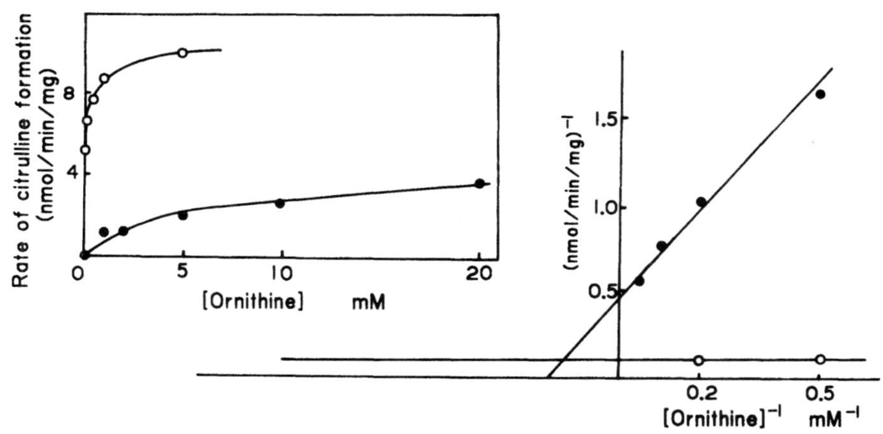

Fig. 7. Effect of ornithine concentration on citrulline synthesis in mitochondria isolated from the liver of a patient with hyperornithinemia, hyperammonemia, and homocitrullinuria (●) and a control subject (○)

times revealed a decreased or undetectable incorporation of ornithine into the isolated mitochondria. Although the method used seems to be not very accurate and reliable, it is the first indication of a defective ornithine transport system in the liver mitochondria of HHH.

The fundamental research on mitochondrial ornithine transport system has been performed mainly with rat liver mitochondria by Gamble and Lehninger (1973), Bryla and Harris (1976), McGivan et al. (1977), and Hommes et al. (1983b). Bradford and McGivan (1980) suggested the existence of two distinct systems for ornithine transport: one exchanges ornithine for citrulline, functioning as a part of the urea cycle, and the other transports ornithine in exchange for H^+, controlling the rate of ornithine degradation by ornithine transaminase. But the relations between the two systems in the hepatocytes and fibroblasts, and between the systems and the defect in HHH, are quite unclear. Research on the protein operating on ornithine transport is necessary for further understanding of HHH.

10 Conclusions

Since H.A. Krebs, many studies have been carried out on the urea cycle, its enzymes, and their deficiencies in man. This paper has reviewed biochemical and molecular research on this subject to date. By now, all of the urea cycle enzymes other than *N*-acetylglutamate synthetase have been purified to homogeneity from human liver and characterized in detail. Clones of cDNA for all the enzymes have been isolated. Primary amino acid sequences of OTC, ASS, and arginase have been deduced from their cDNA nucleotide sequences. RFLPs were found in the gene loci of CPS, OTC, and ASS.

These results may enable us to clarify the pathogenesis of urea cycle enzyme deficiencies at the molecular level, and may offer new therapeutic procedures, including gene therapy. Research on the organ-specific enzyme defects seen in citrullinemia, argininosuccinic aciduria, and probably argininemia may give us some insights into organ-specific gene expression in general.

On the other hand, the characterization of membrane transport proteins is still in the early stages. Purification of the membrane proteins may be necessary for further research.

Acknowledgements. This work was partially supported by the Specific Project Research Grant 60127010 (Inborn Error of Metabolism) from the Ministry of Education, Science and Culture of Japan.

References

Adcock MW, O'Brien WE (1984) Molecular cloning of cDNA for rat and human carbamyl phosphate synthetase I. J Biol Chem 259:13471–13476
Akaboshi I, Endo F, Matsuda I, Saheki T (1983) Kinetic analysis of argininosuccinate synthetase in a variant form of citrullinemia. J Inherited Metab Dis 6:36–39
Akaboshi I, Endo F, Hayashida S, Matsuda I (submitted) A case of argininosuccinic aciduria (in Japanese). J Jpn Pediatr Soc
Allan JD, Cusworth DC, Dent CE, Wilson VK (1958) A disease, probably hereditary, characterised by severe mental deficiency and a constant gross abnormality of amino acid metabolism. Lancet i:182–187
Amaya Y, Kawamoto S, Oda T, Kusumi T, Saheki T, Kimura S, Mori M (1986) Molecular cloning of cDNA for argininosuccinate lyase of rat liver. Biochem Int 13:433–438
Arashima S, Matsuda I (1972) A case of carbamyl phosphate synthetase deficiency. Tohoku J Exp Med 107:143–147
Argan C, Shore GC (1985) The precursor to ornithine carbamoyltransferase is transported to mitochondria as a 5 S complex containing an import factor. Biochem Biophys Res Commun 131:289–298
Armstrong MD, Robinow M (1967) A case of hyperlysinemia: biochemical and clinical observations. Pediatrics 39:546–554
Bachmann C, Colombo JP (1980) Diagnostic value of orotic acid excretion in heritable disorders of the urea cycle and in hyperammonemia due to organic acidurias. Eur J Pediatr 134:109–113
Bachmann C, Colombo JP (1983) Increased tryptophan uptake into the brain in hyperammonemia. Life Sci 33:2417–2424
Bachmann C, Colombo JP (1984) Increase of tryptophan and 5-hydroxyindole acetic acid in the brain of ornithine carbamoyltransferase deficient sparse-fur mice. Pediatr Res 18:372–375
Bachmann C, Krahenbuhl S, Colombo JP, Schubiger G, Jaggi KH, Tonz O (1981) *N*-Acetylglutamate synthetase deficiency: a disorder of ammonia detoxication. N Engl J Med 304:543–544
Bachmann C, Krahenbuhl S, Colombo JP (1982) Purification and properties of acetyl-CoA:L-glutamate *N*-acetyltransferase from human liver. Biochem J 205:123–127
Bascur L, Cabello J, Veliz M, Gonzalez A (1966) Molecular forms of human-liver arginase. Biochim Biophys Acta 128:149–154
Beaudet AL, O'Brien WE, Bock HGO, Freytag SO, Su TS (1986) The human argininosuccinate synthetase locus and citrullinemia. In: Harris H, Hirschhorn K (eds) Advances in human genetics vol 15. Plenum, New York, pp 161–196
Bernar J, Hanson RA, Kern R, Phoenix B, Shaw KNF, Cederbaum SD (1986) Arginase deficiency in a 12-year-old boy with mild impairment of intellectual function. J Pediatr 108:432–435
Berüter J, Colombo JP, Bachmann C (1978) Purification and properties of arginase from human liver and erythrocytes. Biochem J 174:449–454
Bock HGO, Su TS, O'Brien WE, Beaudet AL (1983) Sequence for human argininosuccinate synthetase cDNA. Nucleic Acids Res 11:6505–6512
Bradford NM, McGivan JD (1980) Evidence for the existence of an ornithine/citrulline antiporter in rat liver mitochondria. FEBS Lett 113:294–298
Bray RC, Ratner S (1971) Argininosuccinase from bovine kidney: Comparison of catalytic, physical, and chemical properties with the enzyme from bovine liver. Arch Biochem Biophys 146:531–541
Briand P, Cathelineau L, Kamoun P, Gigot D, Penninckx M (1981) Increase of ornithine transcarbamylase protein in sparse-fur mice with ornithine transcarbamylase deficiency. FEBS Lett 130:65–68

Briand P, Francois B, Rabier D, Cathelineau L (1982) Ornithine transcarbamylase deficiencies in human males. Kinetic and immunochemical classification. Biochim Biophys Acta 704:100–106

Briand P, Miura S, Mori M, Cathelineau L, Kamoun P, Tatibana M (1983) Cell-free synthesis and transport of precursors of mutant ornithine carbamoyltransferases into mitochondria. Biochim Biophys Acta 760:389–397

Brusdeilins M, Kuhner R, Schumacher K (1985) Purification, affinity to anti-human arginase immunoglobulin-Sepharose 4B and subunit molecular weights of mammalian arginases. Biochim Biophys Acta 840:79–90

Bryla J, Harris EJ (1976) Accumulation of ornithine and citrulline in rat liver mitochondria in relation to citrulline formation. FEBS Lett 72:331–336

Buist NRM, Kennaway NG, Hepburn CA, Strandholm JJ, Ramberg DA (1974) Citrullinemia: investigation and treatment over a four-year period. J Pediatr 85:208–214

Carritt B (1977) Somatic cell genetic evidence for the presence of a gene for citrullinemia on human chromosome 9. Cytogenet Cell Genet 19:44–48

Carson NAJ, Scally BG, Neill DW, Carré IJ (1968) Saccharopinuria: a new inborn error of lysine metabolism. Nature 218:679

Carter AL, Eller AG, Rufo S, Metoki K, Hommes FA (1984) Further evidence for a separate enzymic entity for the synthesis of homocitrulline, distinct from the regular ornithine transcarbamylase. Enzyme 32:26–36

Carton D, De Schrijver F, Kint J, Van Durme J, Hooft C (1969) Case report. Argininosuccinic aciduria. Neonatal variant with rapid fatal course. Acta Paediatr Scand 58:528–534

Carvajal N, Cederbaum SD (1986) Kinetics of inhibition of rat liver and kidney arginases by proline and branched-chain amino acids. Biochim Biophys Acta 870:181–184

Carvajal N, Venegas A, Oestreicher G, Plaza M (1971) Effect of manganese on the quaternary structure of human liver arginase. Biochim Biophys Acta 250: 437–442

Cathelineau L, Petit FP, Coude FX, Kamoun PP (1979) Effect of propionate and pyruvate on citrulline synthesis and ATP content in rat liver mitochondria. Biochem Biophys Res Commun 90:327–332

Cathelineau L, Dinh DP, Boue J, Saudubray JM, Farriaux JP, Kamoun P (1981a) Improved method for the antenatal diagnosis of citrullinemia. Clin Chim Acta 116:111–115

Cathelineau L, Dinh DP, Briand P, Kamoun P (1981b) Studies on complementation in argininosuccinate synthetase and argininosuccinate lyase deficiencies in human fibroblasts. Hum Genet 57:282–284

Cederbaum SD, Shaw KNF, Dancis J, Hutzler J, Blaskovics JC (1979a) Hyperlysinemia with saccharopinuria due to combined lysine-ketoglutarate reductase and saccharopine dehydrogenase deficiencies presenting as cystinuria. J Pediatr 95:234–238

Cederbaum SD, Shaw KNF, Spector EB, Verity MA, Snodgrass PJ, Sugarman GI (1979b) Hyperargininemia with arginase deficiency. Pediatr Res 13:827–833

Chen PC, Broome JD (1980) Mouse macrophage arginase (40777). Proc Soc Exp Biol Med 163:354–359

Christensen E, Brandt NJ, Philip J, Kennaway NG (1980) Citrullinaemia: the possibility of prenatal diagnosis. J Inherited Metab Dis 3:73–75

Christensen E, Brandt NJ, Philip J, Bang J, (1985) Exclusion of citrullinaemia in the first trimester of pregnancy by direct assay of argininosuccinate synthetase in chorionic villi. Prenat Diagn 5:299–301

Conboy JG, Kalousek F, Rosenberg LE (1979) In vitro synthesis of a putative precursor of mitochondrial ornithine transcarbamylase. Proc Natl Acad Sci USA 76:5724–5727

Corredor C, Brendel K, Bressler R (1967) Studies on the mechanism of the hypoglycemic action of 4-pentenoic acid. Proc Natl Acad Sci USA 58:2299–2306

Coude FX, Sweetman L, Nyhan WL (1979) Inhibition by propionyl-Coenzyme A of *N*-acetylglutamate synthetase in rat liver mitochondria. A possible explanation for hyperammonemia in propionic and methylmalonic acidemia. J Clin Invest 64:1544–1551

Coude FX, Ogier H, Marsac C, Munnich A, Charpentier C, Saudubray JM (1981) Secondary citrullinemia with hyperammonemia in four neonatal cases of pyruvate carboxylase deficiency. Pediatrics 68:914

Coude FX, Ogier H, Grimber G, Parvy Ph, Pham Dihn D, Charpentier C, Saudubray JM (1982) Correlation between blood ammonia concentration and organic acid accumulation in isovaleric and propionic acidemia. Pediatrics 69:115–117

Coude FX, Grimber G, Parvy P, Rabier D (1983) Role of *N*-acetylglutamate and acetyl-CoA in the inhibition of ureagenesis by isovaleric acid in isolated rat hepatocytes. Biochim Biophys Acta 761:13–16

De Groot CJ, van Zonneveld AJ, Mooren PG, Zonneveld D, van den Dool A, van den Bogaert AJW, Lamers WH, Moorman AFM, Charles R (1984) Regulation of mRNA levels of rat liver carbamoylphosphate synthetase by glucocorticoids and cyclic AMP as estimated with a specific cDNA. Biochem Biophys Res Commun 124:882–888

De Martinville B, Kunkel LM, Bruns G, Morlé F, Koenig M, Mandel JL, Horwich A, Latt SA, Gusella JF, Housman D, Francke U (1985) Localization of DNA sequences in region Xp21 of the human X chromosome: Search for molecular markers close to the Duchenne muscular dystrophy locus. Am J Hum Genet 37:235–249

Desjeux J-F, Rajantie J, Simell O, Dumontier A-M, Perheentupa J (1980) Lysine fluxes across the jejunal epithelium in lysinuric protein intolerance. J Clin Invest 65:1382–1387

Devaney M, Powers-Lee SG (1984) Immunological cross-reactivity between carbamyl phosphate synthetases I, II, and III. J Biol Chem 259:703–706

Dizikes GJ, Cederbaum SD (1984) Isolation and characterization of cDNA clones for rat liver arginase. Am J Hum Genet 36:135S

Dizikes GJ, Grody WW, Kern RM, Cederbaum SD (1985) Isolation and study of cDNA clones for human liver arginase. Am J Hum Genet 37:A152

Endres W, Schaller R, Shin YS (1984) Diagnosis and treatment of argininaemia. Characteristics of arginase in human erythrocytes and tissues. J Inherited Metab Dis 7:8

Farriaux JP, Dhondt JL, Formstecher P, Martin JJ, Pollitt RJ, Kint J, Lagrou A, Mardens Y, Fontaine G (1976) Pathological and biochemical studies on a neonatal case of argininosuccinic aciduria. Acta Neurol Belg 76:26–34

Fearon ER, Mallonee RL, Phillips AP III, O'Brien WE, Brusilow SW, Adcock MW, Kirby LT (1985) Genetic analysis of carbamyl phosphate synthetase I deficiency. Hum Genet 70:207–210

Fell V, Pollitt RJ, Sampson GA, Wright T (1974) Ornithinemia, hyperammonemia, and homocitrullinuria. A disease associated with mental retardation and possibly caused by defective mitochondrial transport. Am J Dis Child 127:752–756

Fleisher LD, Harris CJ, Mitchell DA, Nadler HL (1983) Citrullinemia: prenatal diagnosis of an affected fetus. Am J Hum Genet 35:85–90

Forget BG, Benz EJ, Weisman SM (1983) Normal human globin gene structure and mutations causing the β-thalassemia syndromes. In: Caskey CT, White RL (eds) Recombinant DNA applications to human disease. Cold Spring Harbor Laboratory, New York, pp 3–17 (Banbury reports vol 14)

Freeman JM, Nicholson JF, Schimke RT, Rowland LP, Carter S (1970) Congenital hyperammonemia. Association with hyperglycinemia and decreased levels of carbamylphosphate synthetase. Arch Neurol 23:430–437

Freytag SO, Beaudet AL, Bock HGO, O'Brien WE (1984a) Molecular structure of the human argininosuccinate synthetase gene: Occurrence of alternative mRNA splicing. Mol Cell Biol 4:1978–1984

Freytag SO, Bock HGO, Beaudet AL, O'Brien WE (1984b) Molecular structures of human argininosuccinate synthetase pseudogenes: evolutionary and mechanistic implications. J Biol Chem 259:3160–3166

Funahashi M, Kato H, Shiosaka S, Nakagawa H (1981) Formation of arginine and guanidinoacetic acid in the kidney in vivo. Their relations with the liver and their regulation. J Biochem 89:1347–1356

Gaasbeek Janzen JW, Moorman AFM, Lamers WH, Los JA, Charles R (1981) The localization of carbamyl-phosphate synthase in adult rat liver. Biochem Soc Trans 9:279

Gamble JG, Lehninger AL (1973) Transport of ornithine and citrulline across the mitochondrial membrane. J Biol Chem 248:610–618

Gatfield PD, Taller E, Wolfe DM, Haust MD (1975) Hyperornithinemia, hyperammonemia, and homocitrullinuria associated with decreased carbamyl phosphate synthetase I activity. Pediatr Res 9:488–497

Gebhardt R, Mecke D (1983) Heterogeneous distribution of glutamine synthetase among rat liver parenchymal cells in situ and in primary culture. EMBO J 2: 567–570

Glick NR, Snodgrass PJ, Schafer IA (1976) Neonatal argininosuccinic aciduria with normal brain and kidney but absent liver argininosuccinate lyase activity. Am J Hum Genet 28:22–30

Goldfischer S, Moore CL, Johnson AB, Spiro AJ, Valsamis MP, Wisniewski HK, Ritch RH, Norton WT, Rapin I, Gartner LM (1973) Peroxisomal and mitochondrial defects in the cerebro-hepato-renal syndrome. Science 182:62–64

Goto Y, Habara S, Nagai H, Kida K, Matsuda H (1983) Sibling case with carbamylphosphate synthetase deficiency (in Japanese). Abstract of Japanese Society for Inherited Metabolic Disease, pp 40–41

Gray RGF, Hill SE, Pollitt RJ (1982) Reduced ornithine catabolism in cultured fibroblasts and phytohaemagglutinin-stimulated lymphocytes from a patient with hyperornithinaemia, hyperammonaemia and homocitrullinuria. Clin Chim Acta 118:141–148

Gray RGF, Hill SE, Pollitt RJ (1983) Studies on the pathway from ornithine to proline in cultured skin fibroblasts with reference to the defect in hyperornithinaemia with hyperammonaemia and homocitrullinuria. J Inherited Metab Dis 6:143–148

Greenberg DM, Bagot AE, Roholt OA (1956) Liver arginase. III Properties of highly purified arginase. Arch Biochem Biophys 62:446–453

Grisolia S, Baguena R, Mayor F (eds) (1976) The urea cycle. Wiley, New York

Gruskay JA, Rosenberg LE (1979) Inhibition of hepatic mitochondrial carbamylphosphate synthetase (CPSI) by acyl CoA esters: possible mechanism of hyperammonemia in the organic acidemias. Pediatr Res 13:475

Guthöhrlein G, Knappe J (1968) Structure and function of carbamoylphosphate synthetase. 1. Transitions between twocatalytically inactive forms and the active form. Eur J Biochem 7:119–127

Hata A, Tsuzuki T, Shimada K, Takiguchi M, Mori M, Matsuda I (1986) Isolation and characterization of the human ornithine transcarbamylase gene: structure of the 5′-end region. J Biochem 100:717–729

Häussinger D (1983) Hepatocyte heterogeneity in glutamine and ammonia metabolism and the role of an intercellular glutamine cycle during ureogenesis in perfused rat liver. Eur J Biochem 133:269–275

Haust MD, Gatfield PD, Gordon BA (1981) Ultrastructure of hepatic mitochondria in a child with hyperornithinemia, hyperammonemia and homocitrullinuria. Hum Pathol 12:212–213

Havir EA, Tamir H, Ratner S, Warner RC (1965) Biosynthesis of urea. XI. Preparation and properties of crystalline argininosuccinase. J Biol Chem 240: 3079–3088

Herzfeld A, Raper SM (1976) The heterogeneity of arginases in rat tissues. Biochem J 153:469–478

Hirsch-Kolb H, Greenberg DM (1968) Molecular characteristics of rat liver arginase. J Biol Chem 243:6123–6129

Holland PC, Sherratt HSA (1973) Biochemical effects of the hypoglycaemic compound pent-4-enoic acid and related non-hypoglycaemic fatty acids. Biochem J 136:157–171

Holzgreve W, Golbus MS (1984) Prenatal diagnosis of ornithine transcarbamylase deficiency utilizing fetal liver biopsy. Am J Hum Genet 36:320–328

Hommes FA, Ho CK, Roesel RA, Coryell ME, Gordon BA (1982) Decreased transport of ornithine across the inner mitochondrial membrane as a cause of hyperornithinaemia. J Inherited Metab Dis 5:41–47

Hommes FA, Eller AG, Scott DF, Carter AL (1983a) Separation of ornithine and lysine activities of the ornithine-transcarbamylase-catalyzed reaction. Enzyme 29:271–277

Hommes FA, Kitchings L, Eller AG (1983b) The uptake of ornithine and lysine by rat liver mitochondria. Biochem Med 30:313–321

Hoogenraad NJ, Sutherland TM, Howlett GJ (1980) Purification of ornithine transcarbamylase from rat liver by affinity chromatography with immobilized transition-state analog. Anal Biochem 101:97–102

Horwich AL, Kraus JP, Williams K, Kalousek F, Koningsberg W, Rosenberg LE (1983) Molecular cloning of the cDNA for rat ornithine transcarbamylase. Proc Natl Acad Sci USA 80:4258–4262

Horwich AL, Fenton WA, Williams KR, Kalousek F, Kraus JP, Doolittle RF, Koningsberg W, Rosenberg LE (1984) Structure and expression of a complementary DNA for the nuclear coded precursor of human mitochondrial ornithine transcarbamylase. Science 224:1068–1074

Horwich AL, Kalousek F, Rosenberg LE (1985) Arginine in the leader peptide is required for both import and proteolytic cleavage of a mitochondrial precursor. Proc Natl Acad Sci USA 82:4930–4933

Igarashi N, Kubo M, Ohki T, Suzuki Y, Okabe I (1986) An infantile case of Zellweger syndrome with deficiencies of peroxisomal β-oxidation enzymes (in Japanese). J Jpn Pediatr Soc 90:851–859

Imamura Y, Kobayashi K, Yamashita T, Saheki T, Ichiki H, Hashida S, Ishikawa E (1987) Clinical application of enzyme immunoassay in the analysis of citrullinemia. Clin Chim Acta 164:201–208

Jacoby LB, Shih VE, Struckmeyer C, Niermeijer MF, Boue J (1981) Variation in argininosuccinate synthetase activity in amniotic fluid cell cultures: implications for prenatal diagnosis of citrullinemia. Clin Chim Acta 116:1–7

James JH, Ziparo V, Jeppsson B, Fischer JE (1979) Hyperammonaemia, plasma aminoacid imbalance, and blood-brain amino acid transport: a unified theory of portal-systemic encephalopathy. Lancet 2:772–775

Jinno Y, Nomiyama H, Wakasugi S, Shimada K, Matsuda I, Saheki T (1984) Isolation and characterization of phage clones carrying the human argininosuccinate synthetase-like genes. J Inherited Metab Dis 7:133–134

Jinno Y, Matuo S, Nomiyama H, Shimada K, Matsuda I (1985) Novel structure of the 5′ end region of the human argininosuccinate synthetase. J Biochem 98: 1395–1403

Jinno Y, Niikawa N, Shimada K, Matsuda I (1986) Restriction fragment length polymorphisms in the 5′ end region of the human argininosuccinate synthetase gene. J Inherited Metab Dis 9:317–320

Jonung T, Rigotti P, Jeppsson B, James JH, Peters JC, Fischer JE (1984) Methionine sulfoximine prevents the accumulation of large neutral amino acids in brain of hyperammonemic rats. J Surg Res 36:349–353

Kakinuma H, Ohtake A, Ogura N, Takayanagi M, Nakajima H, Nishioka T, Matsuura Y, Takeuchi Y, Asanuma K (1984) Two siblings with complete carbamyl phosphate synthetase I deficiency. Acta Paediatr Jpn 26:16–19

Kalousek F, Francois B, Rosenberg LE (1978) Isolation and characterization of ornithine transcarbamylase from normal human liver. J Biol Chem 253:3939–3944

Kalumuck KE, McInnes RR, Adcock MW, Beaudet AL, O'Brien WE (1985) Cloning of a full length cDNA for human argininosuccinate lyase. Am J Hum Genet 37:A159

Kamimura N, Mito T, Nakano C, Ishii S, Nakao Y, Kisa T (1986) A case with argininosuccinic aciduria (in Japanese). Jpn Soc Inherited Metab Dis 2:78

Kamoun P, Parvy PH, Phan Dinh D, Boue J, Cathelineau L (1983) Citrulline in aminotic fluid and the prenatal diagnosis of citrullinemia. Prenat Diagn 3:53–56

Kang SS, Wong PWK, Melyn MA (1983) Hyperargininemia: effect of ornithine and lysine supplementation. J Pediatr 103:763–765

Kawamoto S, Ishida H, Mori M, Tatibana M (1982) Regulation of *N*-acetylglutamate synthetase in mouse liver. Postprandial changes in sensitivity to activation by arginine. Eur J Biochem 123:637–641

Kawamoto S, Amaya Y, Oda T, Kusumi T, Saheki T, Kimura S, Mori M (1985) Molecular cloning of cDNAs for arginase and argininosuccinate lyase of rat liver. Jpn Soc Inherited Metab Dis 2:83

Kawamoto S, Amaya Y, Oda T, Kusumi T, Saheki T, Kimura S, Mori M (1986) Cloning and expression in ***Escherichia coli*** of cDNA for arginase of rat liver. Biochem Biophys Res Commun 136:955–961

Kawamoto S, Sonoda T, Ohtake A, Suzuki Y, Tatibana M (submitted) Alteration in arginine activation of *N*-acetylglutamate synthetase in vitro by disulfide or thiol compounds

Kaysen GA, Strecker HJ (1973) Purification and properties of arginase of rat kidney. Biochem J 133:779–788

Kennaway NG, Harwood PJ, Ramberg DA, Koler RD, Buist NRM (1975) Citrullinemia: enzymatic evidence for genetic heterogeneity. Pediatr Res 9:554–558

Kimball ME, Jacoby LB (1980) Purification and properties of argininosuccinate synthetase from normal and canavanine-resistant human lymphoblasts. Biochemistry 19:705–709

Kitajima I, Goto K, Umehara F, Nagamatsu K, Kanehisa Y (1986) A case of lysinuric protein intolerance with intermittent stupor looking like psychomotor seizure in adulthood. Clin Neurol 26:592–600

Kleijer WJ, Blom W, Huijmans JGM, Mooyman MCT, Berger R, Niermeijer MF (1984) Prenatal diagnosis of citrullinemia: elevated levels of citrulline in the amniotic fluid in the three affected pregnancies. Prenat Diagn 4:113–118

Kobayashi K, Itakura Y, Saheki T, Nakano K, Sase M, Oyanagi K, Okamoto R, Mino M (1986a) Absence of argininosuccinate lyase protein in the liver of two patients with argininosuccinic aciduria. Clin Chim Acta 159:59–67

Kobayashi K, Saheki T, Imamura Y, Noda T, Inoue I, Matuo S, Hagihara S, Nomiyama H, Jinno Y, Shimada K (1986b) Messenger RNA coding for argininosuccinate synthetase in citrullinemia. Am J Hum Genet 38:667–680

Kodama H, Samukawa K, Okada S, Nose O, Maki I, Yamaguchi M, Yabuuchi H (1983) Study of ammonia metabolism in a patient with ornithine transcarbamylase deficiency using a ^{15}N tracer. Clin Chim Acta 132:267–275

Kodama H, Ohtake A, Mori M, Okabe I, Tatibana M, Kamoshita S (1986) Ornithine transcarbamylase deficiency. A case with a truncated enzyme precursor and a case with undetectable mRNA activity. J Inherited Metab Dis 9:175–185

Konarska L, Tomaszewski L (1986) A simple quantitative micromethod of arginase assay in blood spots dried on filter paper. Clin Chim Acta 154:7–18

Konarska L, Wiesmann U, Colombo JP (1981) Arginase activity in human fibroblast cultures. Clin Chim Acta 115:85–92

Konarska L, Wiesmann U, v. Fellenberg R, Colombo JP (1983) Isoenzyme pattern and immunological properties of arginase in normal and hyperargininemia fibroblasts. Enzyme 29:44–53

Konarska L, Tomaszewski L, Colombo JP, Terheggen HG (1985) Human salivary arginase and its deficiency in arginaemia. J Clin Chem Clin Biochem 23:337–342

Kooka T, Higashi Y, Uebayashi Y, Kobayashi R (1977) A special form of hepato-cerebral degeneration with citrullinemia. Neurol Med 6:47–53

Lambert M, Simard L, Ray P, McInnes RR (1985) Cloning of argininosuccinate lyase (ASL) and molecular analysis of ASL expression in rat hepatoma lines. Am J Hum Genet 37:A163

Lambert MA, Simard LR, Ray PN, McInnes RR (1986) Molecular cloning of cDNA for rat argininosuccinate lyase and its expression in rat hepatoma cell lines. Mol Cell Biol 6:1722–1728

Levin B, Dobbs RH, Burgess EA, Palmer T (1969) Hyperammonaemia. A variant type of deficiency of liver ornithine transcarbamylase. Arch Dis Child 44:162–169

Levin B, Oberholzer VG, Palmer T (1973) Citrullinemia and an alternative urea cycle. Pediatr Res 7:728

Lindgren V, de Martinville B, Horwich AL, Rosenberg LE, Francke U (1984) Human ornithine transcarbamylase locus mapped to band Xp21.1 near the Duchenne muscular dystrophy locus. Science 226:698–700

Lockridge O, Spector EB, Bloom AD (1977) Argininosuccinate synthetase activity in cultured human lymphocytes. Biochem Genet 15:395–407

Lusty CJ (1978) Carbamylphosphate synthetase I of rat-liver mitochondria. Purification, properties, and polypeptide molecular weight. Eur J Biochem 85:373–383

Lusty CJ, Jilka RL, Nietsch EH (1979) Ornithine transcarbamylase of rat liver. Kinetic, physical, and chemical properties. J Biol Chem 254:10030–10036

Mans AM, Biebuyck JF, Hawkins RA (1983) Ammonia selectively stimulates neutral amino acid transport across blood-brain barrier. Am J Physiol 245:C74–77

Marescau B, Pintens J, Lowenthal A, Terheggen HG, Adriaenssens K (1979) Arginase and free amino acids in hyperargininemia: leucocyte arginase as a diagnostic parameter for heterozygotes. J Clin Chem Clin Biochem 17:211–217

Marescau B, Qureshi IA, De Deyn P, Letarte J, Ryba R, Lowenthal A (1985) Guanidino compounds in plasma, urine and cerebrospinal fluid of hyperargininemic patients during therapy. Clin Chim Acta 146:21–27

Marshall M, Cohen PP (1972) Ornithine transcarbamylase from ***Streptococcus faecalis*** and bovine liver. I. Isolation and subunit structure. J Biol Chem 247: 1641–1653

Matsuda I, Arashima S, Imanishi Y, Yamamoto J, Akaboshi I, Shinozuka S, Nagata N (1979) Lysine intolerance in a variant form of citrullinemia. Pediatr Res 13: 1134–1136

Matsuo M, Ookita K, Hashimoto T, Takemine H, Ito H, Koike K, Koike M (1982) A case with deficiency of pyruvate decarboxylase accompanied with hyper-ammonemia (in Japanese). J Jpn Pediatr Soc 86:214–219

McGivan JD, Bradford NM, Mendes-Mourao J (1976) The regulation of carbamoyl-phosphate synthase activity in rat liver mitochondria. Biochem J 154:415–421

McGivan JD, Bradford NM, Beavis AD (1977) Factors influencing the activity of ornithine aminotransferase in isolated rat liver mitochondria. Biochem J 162: 147–156

McInnes RR, Shih V, Chilton S (1984) Interallelic complementation in an inborn error of metabolism: genetic heterogeneity in argininosuccinate lyase deficiency. Proc Natl Acad Sci USA 81:4480–4484

McMurray WC, Mohyuddin F, Rossiter RM, Rathbun JC, Valentine GH, Koegler SJ, Zarfas DE (1962) Citrullinuria. A new aminoaciduria associated with mental retardation. Lancet i:138

Metoki K, Hommes FA, Dyken P, Kelloes C, Trefz J (1984) Ultrastructural changes in fibroblast mitochondria of a patient with HHH-syndrome. J Inherited Metab Dis 7:147–150

Mezl VA, Knox WE (1977) Metabolism of arginine in lactating rat mammary gland. Biochem J 164:105–113

Michels VV, Beaudet AL (1978) Arginase deficiency in multiple tissues in argininemia. Clin Genet 13:61–67

Mizutani A (1968) Cytochemical demonstration of ornithine carbamoyltransferase activity in liver mitochondria of rat and mouse. J Histochem Cytochem 16: 172–180

Mori M, Morita T, Ikeda F, Amaya Y, Tatibana M, Cohen PP (1981) Synthesis, intracellular transport, and processing of the precursors for mitochondrial ornithine transcarbamylase and carbamylphosphate synthetase I in isolated hepatocytes. Proc Natl Acad Sci USA 78:6056–6060

Murakami-Murofushi K, Ratner S (1979) Argininosuccinase from bovine brain: isolation and comparison of catalytic, physical, and chemical properties with the enzymes from liver and kidney. Anal Biochem 95:139–155

Naylor SL, Klebe RJ, Shows TB (1978) Argininosuccinic aciduria: Assignment of the argininosuccinate lyase gene to the pter →q22 region of human chromosome 7 by bioautography. Proc Natl Acad Sci USA 75:6159–6162

Nomiyama H, Obaru K, Jinno Y, Matsuda I, Shimada K, Miyata T (1986) Amplification of human argininosuccinate synthetase pseudogenes. J Mol Biol 192:221–233

Nussbaum RL, Boggs BA, Beaudet AL, Doyle S, Potter JL, O'Brien WE (1986) New mutation and prenatal diagnosis in ornithine transcarbamylase deficiency. Am J Hum Genet 38:149–158

O'Brien WE (1979) Isolation and characterization of argininosuccinate synthetase from human liver. Biochemistry 18:5353–5356

O'Brien WE, Barr RH (1981) Argininosuccinate lyase: purification and characterization from human liver. Biochemistry 20:2056–2060

Ohtake A, Miura S, Mori M, Takayanagi M, Kakinuma H, Tatibana M, Nakajima H (1984) A carbamyl phosphate synthetase I deficiency with no detectable messenger RNA activity. Acta Paediatr Jpn 26:262–265

Orkin SH (1983) A review of β-thalassemias: The spectrum of gene mutations. In: Caskey CT, White RL (eds) Recombinant DNA applications to human disease. Cold Spring Harbor Laboratory, New York, pp 19–28 (Banbury reports vol 14)

Oyanagi K, Tsuchiyama A, Itakura Y, Sogawa H, Wagatsuma K, Nakao T (1983) The mechanism of hyperammonaemia and hyperornithinaemia in the syndrome of hyperornithinaemia, hyperammonaemia with homocitrullinuria. J Inherited Metab Dis 6:133–134

Oyanagi K, Itakura Y, Tsuchiyama A, Uetsuji N, Tsugawa S, Sase M, Saheki T (1985) Argininosuccinic aciduria associated with argininosuccinate synthetase deficiency in liver (in Japanese). Rinsho Shoni Igaku 33:347–352

Palekar AG, Mantagos S (1981) Human liver argininosuccinase purification and partial characterization. J Biol Chem 256:9192–9194

Perry TL, Wirtz MLK, Kennaway NG, Hsia YE, Atienza FC, Uemura HS (1980) Amino acid and enzyme studies of brain and other tissues in an infant with argininosuccinic aciduria. Clin Chim Acta 105:257–267

Pierson DL, Brien JM (1980) Human carbamylphosphate synthetase I. Stabilization, purification, and partial characterization of the enzyme from human liver. J Biol Chem 255:7891–7895

Pierson DL, Cox SL, Gilbert BE (1977) Human ornithine transcarbamylase. Purification and characterization of the enzyme from normal liver and the liver of a Reye's syndrome patient. J Biol Chem 252:6464–6469
Pohjanpelto P, Holtta E (1983) Arginase activity of different cells in tissue culture. Biochim Biophys Acta 757:191–195
Pollitt RJ (1973) Argininosuccinate lyase levels in blood, liver and cultured fibroblasts of a patient with argininosuccinic aciduria. Clin Chim Acta 46:33–37
Pollock MA, Cumberbatch M, Bennett MJ, Gray RGF, Brand M, Hyland K, Congdon PJ, Pitts-Tucker T, Gray S (1986) Pyruvate carboxylase deficiency in twins. J Inherited Metab Dis 9:29–30
Powers SG (1981) Regulation of rat liver carbamyl phosphate synthetase I. Inhibition by metal ions and activation by amino acids and other chelating agents. J Biol Chem 256:11160–11165
Qureshi IA, Letarte J, Ouellet R, Lemieux B (1978) Enzymologic and metabolic studies in two families affected by argininosuccinic aciduria. Pediatr Res 12:256–262
Qureshi IA, Letarte J, Ouellet R, Lelievre M, Laberge C (1981) Ammonia metabolism in a family affected by hyperargininemia. Diabete Metab 7:5–11
Qureshi IA, Letarte J, Ouellet R, Larochelle J, Lemieux B (1983) A new French-Canadian family affected by hyperargininaemia. J Inherited Metab Dis 6:179–182
Raijman L (1979) Double deficiencies of urea cycle enzymes in human liver. Biochem Med 21:226–233
Raijman L, Jones ME (1976) Purification, composition, and some properties of rat liver carbamyl phosphate synthetase (ammonia). Arch Biochem Biophys 175:270–278
Ratner S (1973) Enzymes of arginine and urea synthesis. In: Meister A (ed) Advances in enzymology vol 39. Wiley, New York, pp 1–90
Ratner S (1976) Enzymes of arginine and urea synthesis. In: Grisolia S, Baguena R, Mayor F (eds) The urea cycle. Wiley, New York, pp 181–219
Raushel FM, Seiglie J (1983) Kinetic mechanism of argininosuccinate synthetase. Arch Biochem Biophys 225:979–985
Rigotti P, Jonung T, Peters JC, James JH, Fischer JE (1985) Methionine sulfoximine prevents the accumulation of large neutral amino acids in brain of portacaval-shunted rats. J Neurochem 44:929–933
Rochovansky O, Kodowaki H, Ratner S (1977) Biosynthesis of urea: molecular and regulatory properties of crystalline argininosuccinate synthetase. J Biol Chem 252:5287–5294
Rodeck CH, Patrick AD, Pembrey ME, Tzannatos C, Whitfield AE (1982) Fetal liver biopsy for prenatal diagnosis of ornithine carbamyl transferase deficiency. Lancet ii:297–299
Roerdink FH, Gouw WLM, Okken A, Van der Blij JF, Haan GL Hommes FA, Huisjes HJ (1973) Citrullinemia, report of a case, with studies on antenatal diagnosis. Pediatr Res 7:863–869
Rosenberg LE, Kalousek F, Orsulak MD (1983) Biogenesis of ornithine transcarbamylase in *spf*^{ash} mutant mice: two cytoplasmic precursors, one mitochondrial enzyme. Science 222:426–428
Rozen R, Fox J, Fenton WA, Horwich AL, Rosenberg LE (1985) Gene deletion and restriction fragment length polymorphisms at the human ornithine transcarbamylase locus. Nature 313:815–817
Rubio V, Grisolia S (1981) Human carbamoylphosphate synthetase I. Enzyme 26:233–239
Ryall J, Rauchubinski RA, Nguyen M, Rozen R, Broglie KE, Shore GC (1984) Regulation and expression of carbamyl phosphate synthetase I mRNA in developing rat liver and Morris hepatoma 5123D. J Biol Chem 259:9172–9176

Saheki T, Kusumi T, Takada S, Katsunuma T, Katunuma N (1975) Crystallization and some properties of argininosuccinate synthetase from rat liver. FEBS Lett 58:314–317

Saheki T, Katsunuma T, Sase M (1977a) Changes of ornithine and acetylglutamate concentrations in the livers of rats subjected to dietary transitions. J Biochem 82:551–558

Saheki T, Kusumi T, Takada S, Katsunuma T (1977b) Studies of rat liver argininosuccinate synthetase. I. Physicochemical, catalytic, and immunochemical properties. J Biochem 81:687–696

Saheki T, Ohkubo T, Katsunuma T (1978) Increase in the concentrations of ornithine and acetylglutamate in rat liver in response to urea synthesis stimulated by the injection of an ammonium salt. J Biochem 84:1423–1430

Saheki T, Ueda A, Hosoya M, Kusumi K, Takada S, Tsuda M, Katsunuma T (1981) Qualitative and quantitative abnormalities of argininosuccinate synthetase in citrullinemia. Clin Chim Acta 109:325–335

Saheki T, Ueda A, Hosoya M, Sase M, Nakano K, Katsunuma T (1982a) Enzymatic analysis of citrullinemia (12 cases) in Japan. Adv Exp Med Biol 153:63–76

Saheki T, Ueda A, Iizima K, Yamada N, Kobayashi K, Takahashi K, Katsunuma T (1982b) Argininosuccinate synthetase activity in cultured skin fibroblasts of citrullinemic patients. Clin Chim Acta 118:93–97

Saheki T, Sase M, Nakano K, Azuma F, Katsunuma T (1983a) Some properties of argininosuccinate synthetase purified from human liver and a comparison with the rat liver enzyme. J Biochem 93:1531–1537

Saheki T, Yagi Y, Sase M, Nakano K, Sato E (1983b) Immunohistochemical localization of argininosuccinate synthetase in the liver of control and citrullinemic patients. Biomed Res 4:235–238

Saheki T, Imamura Y, Inoue I, Miura S, Mori M, Ohtake A, Tatibana M, Katsumata N, Ohno T (1984) Molecular basis of ornithine transcarbamylase deficiency lacking enzyme protein. J Inherited Metab Dis 7:2–8

Saheki T, Nakano K, Kobayashi K, Imamura Y, Itakura Y, Sase M, Hagihara S, Matuo S (1985a) Analysis of the enzyme abnormality in eight cases of neonatal and infantile citrullinemia in Japan. J Inherited Metab Dis 8:155–156

Saheki T, Sase M, Nakano K, Yagi Y (1985b) Arginine metabolism in citrullinemic patients. In: Mori A, Cohen BD, Lowenthal A (eds) Guanidines. Plenum, New York, pp 149–158

Saheki T, Kobayashi K, Miura T, Hashimoto S, Ueno Y, Yamasaki T, Araki H, Nara H, Shiozaki Y, Sameshima Y, Suzuki M, Yamauchi Y, Sakazume Y, Akiyama K, Yamamura Y (1986) Serum amino acid pattern of type II citrullinemic patient and effect of oral administration of citrulline. J Clin Biochem Nutr 1:129–142

Sakiyama T, Suzuki T, Owada M, Kitagawa T (1982) First case of argininosuccinic aciduria in Japan: clinical observations and treatment. Adv Exp Med Biol 153: 95–100

Sase M, Kobayashi K, Imamura Y, Saheki T, Nakano K, Miura S, Mori M (1985) Level of translatable messenger RNA coding for argininosuccinate synthetase in the liver of the patients with quantitative-type citrullinemia. Hum Genet 69:130–134

Schimke RT (1964) The importance of both synthesis and degradation in the control of arginase levels in rat liver. J Biol Chem 239:3808–3817

Schutgens RBH, Beemer FA, Tegelaers WHH, De Groot WP (1979) Mild variant of argininosuccinic aciduria. J Inherited Metab Dis 2:13–14

Scott-Emuakpor A, Higgins JV, Kohrman AF (1972) Citrullinemia: a new case, with implications concerning adaptation to defective urea synthesis. Pediatr Res 6:626–633

Shigesada K, Tatibana M (1971a) Enzymatic synthesis of acetylglutamate by mammalian liver preparation and its stimulation by arginine. Biochem Biophys Res Commun 44:1117–1124

Shigesada K, Tatibana M (1971b) Role of acetylglutamate in ureotelism. I. Occurrence and biosynthesis of acetylglutamate in mouse and rat tissues. J Biol Chem 246:5588–5595

Shigesada K, Tatibana M (1978) *N*-Acetylglutamate synthetase from rat-liver mitochondria. Partial purification and catalytic properties. Eur J Biochem 84: 285–291

Shigesada K, Aoyagi K, Tatibana M (1978) Role of acetylglutamate in ureotelism. Variations in acetylglutamate level and its possible significance in control of urea synthesis in mammalian liver. Eur J Biochem 85:385–391

Shih VE (1976) Hereditary urea-cycle disorders. In: Grisolia S, Baguena R, Mayor F (eds) The urea cycle. Wiley, New York, pp 367–414

Shih VE (1978) Urea cycle disorders and other congenital hyperammonemic syndromes. In: Stanbury JB, Wyngaarden JB, Fredrickson DS (eds) The metabolic basis of inherited disease, 4th edn. McGraw-Hill, New York, pp 362–386

Shih VE, Mandell R (1974) Metabolic defect in hyperornithinaemia. Lancet ii: 1522–1523

Shih VE, Efron ML, Moser HW (1969) Hyperornithinemia, hyperammonemia, and homocitrullinuria. A new disorder of amino acid metabolism associated with myoclonic seizures and mental retardation. Am J Dis Child 117:83–92

Shih VE, Mandell R, Herzfeld A (1982) Defective ornithine metabolism in cultured skin fibroblasts from patients with the syndrome of hyperornithinemia, hyperammonemia and homocitrullinuria. Clin Chim Acta 118:149–157

Simell O, Mackenzie S, Clow CL, Scriver CR (1985) Ornithine loading did not prevent induced hyperammonemia in a patient with hyperornithinemia-hyperammonemia-homocitrullinuria syndrome. Pediatr Res 19:1283–1287

Skrzypek-Osiecka I, Robin Y, Porembska Z (1983) Purification of rat kidney arginases A_1 and A_4 and their subcellular distribution. Acta Biochim Pol 30: 83–92

Snyderman SE, Sansaricq C, Chen WJ, Norton PM, Phansalkar SV (1977) Argininemia. J Pediatr 90:563–568

Snyderman SE, Sansaricq C, Norton PM, Goldstein F, (1979) Argininemia treated from birth. J Pediatr 95:61–63

Sonoda T, Tatibana M (1983) Purification of *N*-acetyl-L-glutamate synthetase from rat liver mitochondria and substrate and activator specificity of the enzyme. J Biol Chem 258:9839–9844

Spector EB, Rice SCH, Cederbaum SD (1983) Immunologic studies of arginase in tissues of normal human adult and arginase-deficient patients. Pediat Res 17: 941–944

Stanbury JB, Wyngaarden JB, Frederickson DS (eds) (1978) The metabolic basis of inherited disease, 4th edn. McGraw-Hill, New York

Stewart PM, Walser M (1980) Short term regulation of ureagenesis. J Biol Chem 255:5270–5280

Su TS, Bock HGO, O'Brien WE, Beaudet AL (1981) Cloning of cDNA for argininosuccinate synthetase mRNA and study of enzyme overproduction in a human cell line. J Biol Chem 256:11826–11831

Su TS, Bock HGO, Beaudet AL, O'Brien WE (1982) Molecular analysis of argininosuccinate synthetase deficiency in human fibroblasts. J Clin Invest 70:1334–1339

Su TS, Bock HGO, O'Brien WE, Beaudet AL (1981) Cloning of cDNA for argininosuccinate synthetase mRNA and study of enzyme overproduction in a human

Su TS, Nussbaum RL, Airhart S, Ledbetter DH, Mohandas T, O'Brien WE, Beaudet AL (1984) Human chromosomal assignments for 14 argininosuccinate synthetase pseudogenes: cloned DNAs as reagents for cytogenetic analysis. Am J Hum Genet 36:954–964

Suzuki Y, Orii T, Mori M, Tatibana M, Hashimoto T (1986) Deficient activities and proteins of peroxisomal β-oxidation enzymes in infants with Zellweger syndrome. Clin Chim Acta 156:191–196

Takada S, Kusumi T, Saheki T, Tsuda M, Katsunuma T (1979) Studies of rat liver argininosuccinate synthetase. The presence of three forms, and their physicochemical, catalytic, and immunochemical properties. J Biochem 86:1353–1359

Takiguchi M, Miura S, Mori M, Tatibana M, Nagata S, Kaziro Y (1984) Molecular cloning and nucleotide sequence of cDNA for rat ornithine carbamoyltransferase precursor. Proc Natl Acad Sci USA 81:7412–7416

Terheggen HG, Schwenk A, Lowenthal A, van Sande M, Colombo JP (1969) Argininaemia with arginase deficiency. Lancet ii: 748–749

Terheggen HG, Lowenthal A, Lavinha F, Colombo JP (1975) Familial hyperargininaemia. Arch Dis Child 50:57–62

Van der Heiden C, Gerards LJ, Van Biervliet JPGM, Desplanque J, De Bree PK, Van Sprang FJ, Wadman SK (1976) Lethal neonatal argininosuccinate lyase deficiency in four children from one sibship. Helv Paediatr Acta 31:407–417

Van Elsen AF, Leroy JG (1975) Arginase isoenzymes in human diploid fibroblasts. Biochem Biophys Res Commun 62:191–198

Van Elsen AF, Leroy JG (1977) Human hyperargininemia: a mutation not expressed in skin fibroblasts? Am J Hum Genet 29:350–355

Vidailhet M, Levin B, Dautrevaux M, Paysant P, Gelot S, Badonnel Y, Pierson M, Neimann N (1971) Citrullinemie. Arch Franc Ped 28:521–532

Vielle-Breitburd F, Orth G (1972) Rabbit liver L-arginase. Purification, properties and subunit structure. J Biol Chem 247:1227–1235

Walser M (1983) Urea cycle disorders and other hereditary hyperammonemic syndromes. In: Stanbury JB, Wyngaarden JB, Frederickson DS, Goldstein JL, Brown MS (eds) The metabolic basis of inherited disease. McGraw-Hill, New York, pp 402–438

Windmueller HG, Spaeth AE (1981) Source and fate of circulating citrulline. Am J Physiol 241:E473–E480

Wright T, Pollitt R (1973) Psychomotor retardation, epileptic and stuporous attacks, irritability and ataxia associated with ammonia intoxication, high blood ornithine levels and increased homocitrulline in the urine. Proc R Soc Med 66:221–226

Yajima Y, Hirasawa T, Saheki T (1982) Diurnal fluctuation of blood ammonia levels in adult-type citrullinemia. Tohoku J Exp Med 137:213–220

Yamamoto H, Sawada Y, Hase Y, Fukuda Y, Tsuruhara T, Ohura T, Yamaguchi S, Aono S, Fujimoto A, Ohtake H, Miyagi T (1983) A case with argininosuccinic aciduria discovered by mass-screening test (in Japanese). Abstracts of Japanese Society of Inherited Metabolic Disease, p 62

Yokoi T, Honke K, Funabashi T, Hayashi R, Suzuki Y, Taniguchi N, Hosoya M, Saheki T (1981) Partial ornithine transcarbamylase deficiency simulating Reye syndrome. J Pediatr 99:929–931

Yoshida K, Watanabe M, Kida K, Matsuda H (1984) A case with hyperornithinemia, hyperammonemia and homocitrullinuria (in Japanese). Abstracts of Japanese Society of Inherited Metabolic Disease, p 105

Yoshino M, Kubota K, Yoshida I, Murakami T, Yamashita F (1982) Argininemia: report of a new case and mechanisms of orotic aciduria and hyperammonemia. Adv Exp Med Biol 153:121–125

Rev. Physiol. Biochem. Pharmacol., Vol. 108

Approaches to Visual Cortical Function

RÜDIGER VON DER HEYDT

Contents

Department of Neurology, University Hospital Zurich, CH-8091 Zurich, Switzerland

Introduction

"The way we think of and label the responses of visual cortical cells depends on our underlying assumptions as to how the whole system functions" (MacKay 1985). This explains a general difficulty encountered when trying to survey the studies in this field of physiology. While they all have the use of visual stimuli in common, the concepts or assumptions on which the different investigators have based their studies are sometimes so different that a comparison of the experimental results is hardly possible. Nor is it easy to present them in a logical order. MacKay's remark also highlights the main problem in this whole field of research. Vision is not just imaging and transformation of images – but nobody knows exactly what it is. We study visual cortical processing not knowing what constitutes a step in processing toward the goal.

This article covers only selected areas, roughly circumscribed by the terms spatial filtering, color, and contour, and does not attempt to review the literature completely, not even in these areas. The selection was guided by an interest in two approaches to understanding visual cortical function, each of which has had its impact but still plays only a marginal role with regard to the entire body of physiological studies in the field, while both combined, I anticipate, will be quite successful in the future.

The first is the constructive approach, as exemplified by the work of D. Marr and his group, starting from the question of what has to be computed from an image, and what has to be represented in order to utilize its information efficiently. At first, this appears to be an attempt to create a theoretical visual system out of nothing; but Marr has introduced a tool, the analysis of the physical constraints on images and the corresponding regularities in visual information. To a certain extent these regularities allow a visual system to do the impossible, namely to recover physical properties of visible three-dimensional objects from the two-dimensional intensity arrays present in its eyes. An example that has intrigued many investigators is stereopsis, where the regularity consits of a correlation between the highly similar, but not identical, images in the two eyes, and the physical property derived from it is depth. Physical constraints also play a role in color vision and perception of contour. As a result of the cohesiveness of matter, contours of physical objects are mostly smooth, and due to the diffuseness of illumination, radiance changes across surfaces are mostly gradual. Since it is clear from many examples that biological visual systems make efficient use of this method, the way to understand them must be to identify these regularities, and to invent, or re-invent, the techniques by which the physical object properties could be derived. This has become known as Marr's computational approach (Poggio 1981). In the past, this approach has centered on "early vision" problems (see Poggio and Koch 1985), but in a wider sense, the identification of regularities in visual information and of the way this information co-varies with other sensory signals, might also form a basis for understanding processing at higher stages, including the cognitive level.

It is clear that this approach presents formidable problems. Many of the regularities are not obvious, and which of them should be exploited for a given task and how this could be done is not known. Solutions have often been found by mere guessing, while thorough investigations have encountered deep mathematical problems. Finally, when a suitable algorithm for a task is found, the question remains how it might be implemented. In pursuing the question of what is computed and why, Marr's intuition was often guided by the psychophysics of human vision. The investigation of perception may indeed answer such questions, but the limitations of this method lie in the complexity of the whole, of which the visual system is only a part, and those systems that one might hope to identify are only subsystems. Their outputs are thus not directly observable, and one has to rely on "criteria". A mixture of subsystems with different processing strategies, such as parallel and serial, is characteristic of the visual system (e.g., Sagi and Julesz 1985). Hypothesized intermediate levels of processing, such as the stage of zero-crossing detection, may not be accessible to perception (Marr 1982, p. 75). In short, theoretical intuition needs restraints of experimental facts which psychophysics can only supply in part.

The second approach which tries to relate perception to neural signals could provide these necessary restraints. This approach has been used since it has been possible to register single unit responses to visual stimuli, but after some early attempts that perhaps promised too much, few physiological studies have actually set out in this direction. I am referring here not to the enterprise of drawing parallels between psychophysical and physiological data in general (cf. Jung 1972, 1973), but rather to the specific question: what are perceptions and to which neural events do they correspond? A full understanding of the perception of brightness, color, texture, motion, depth, orientation, contour, shape, etc. would require that we could specifiy in each case a mathematical operator that transforms the images of the two eyes into a quantity that covaries with the given perception over all possible images. We certainly do not know any of these operators, although there are now good approximations for some of them on a restricted set of images.

Indeed, if we had such an understanding of the perception of these qualities, most of the "early vision problems" would be solved. Neurophysiological studies often seem to ignore these questions. In fact, doing single unit recordings I found it difficult to see that there is this basic lack of understanding, and how important these questions are for the investigation of the cortex. I attribute this difficulty to the fact that we usually have simple procedures to hand to measure and specify the physical properties of the laboratory stimuli which give rise to these perceptions. Some of these stimuli are in fact so simple that their perception is "veridical". In order to relate neural signals to perception we first need to design experimental situations in which perception differs markedly from the simple concepts of stimulus properties. A response to the onset of a light, for example, may be related to a change in perception of brightness or to the change in luminance of the stimulus, a selective response to red stimuli could be related to color (the perception) or to the spectral composition of the light, a selective response to edges of a certain orientation to the perception of contour, or to the luminance step, and so on. Early attempts to resolve this ambiguity were the experiments of Baumgartner (1960) and Baumgartner and Hakas (1962) (see also Jung 1972) where it was shown that, with contrast stimuli in which human observers perceive peculiar brightness patterns such as the Mach bands near light-dark gradients or the spots in the Hermann grid, neural signals in the cat visual system correlated with brightness perception rather than the luminance patterns. However the correlations did not hold up in other tests; the neural signals also showed Mach-band-like overshoots in homogenous areas delineated by sharp edges which actually look homogeneous in brightness. Possibly the neural signals correlated with a bandpass-filtered image of the stimulus, i.e., something that lies between the luminance and brightness patterns. Thus, at a second stage we have to design experimental situations which allow us to distinguish between perception and simpler models, such as the spatial filtering concept. Attempts to relate neural signals to color and contour perception will be discussed below.

Two benefits can be expected from such an approach. First, it will lead us to reconsider the psychological concepts of perception (from which the constructive approach has often derived the goals of computation). It will help to define more precisely what is actually computed. Secondly, the site (the cortical area, or the synaptic level) where signals first correlate with a perception, and the way in which they fail to correlate at other (lower) levels, will indicate what is computed before the final goal is achieved, and thus give hints as to how it is achieved.

The reader will notice that these are simplified, programmatic statements. For example, the assumption is made that the flow of information is unidirectional, or "bottom up", as a first approximation. Also, the association between neural activity and perception is always tentative, since correlation may be observed over a limited ensemble of stimuli only, which is usually a tiny subsample of all natural stimuli, or of the stimuli needed for identification of a system in the sense of Wiener.

This review concentrates on studies of single-unit activity, and refers occasionally to psychophysical or theoretical studies. However, it ignores several areas of research. Among these are the fields of visual motion detection and stereopsis, the signal detection approach (exemplified by the work of H.B. Barlow), the attempts to correlate function with morphology and anatomical structure, and the pharmacological approach. The omission of the signal detection approach in particular is regrettable, since several recent studies have come to interesting conclusions about the reliability and detection capabilities of neural responses, which are obviously relevant to both the constructive approach and the correlation with perception. I might defend it by arguing that questions of signal and noise, though certainly very important to perception, are only one aspect of it. By investigating the limits of perception, one can certainly learn much about its mechanisms, bot one might tend to lose sight of the more basic question: what is visual perception?

2 Receptive Fields

In this section I shall discuss some basic questions of receptive field analysis and classification, which will be useful in the following sections, and a few related topics on which evidence seems to be contradictory. With few exceptions, the studies cited here were not aimed at clarifying the visual function of cortical cells, but at elucidating the structure and mechanisms of the receptive fields themselves. They do not, or only incidentally, refer to perception. Since function is our main interest in this review, I shall not review these studies in detail, and often cite a study for a result that may not have been its main concern.

2.1 ON and OFF Responses, Excitation and Inhibition

A few general and simple comments on the terms "excitation and inhibition" and "ON and OFF responses" may be helpful for the following discussion. One thing that I have found confusing in the literature is the frequent use of the terms "ON" and "OFF" as in "ON region" and "OFF region" on the one hand, and "excitatory" and "inhibitory" on the other hand, as if these were interchangeable and there were essentially two functional types of receptive field subregions (in the case of spatial antagonism) or wavelength-dependent effects (in the case of color opponency). Both increments and decrements of light can have effects on the activity of a cell, resulting in either an increase or a decrease of activity.[1] If we consider the possible combinations of these, we have, at this qualitative level of description, six different functions that could be implemented at each point of a receptive field. This may be explained best by a table (Table 1).

For either stimulus event, ON and OFF, the possible effects of increasing (excitatory) or decreasing (inhibitory) the activity are indicated. Of these theoretical types, numbers 1 and 3 would usually be referred to as ON types, and 4 and 6 as OFF types, since they exhibit

Table 1. Elementary responses types

Type	Stimulus	
	ON	OFF
1	e	–
2	–	i
3	e	i
4	–	e
5	i	–
6	i	e

e, excitatory; i, inhibitory

[1] Some authors consequently count excitatory (ON and OFF) and inhibitory (ON and OFF) subregions separately, whether they are spatially coextensive or not. This may lead to confusion if simple cell receptive fields "composed of four to six excitatory regions alternating in space with up to seven inhibitory regions" (Mullikin et al. 1984b) are imagined as fields having up to 13 spatially separate regions, while in fact there were maximally seven such regions.

excitation at light-on and light-off, respectively, although these types are functionally very different. Consider, for example, a unit with a receptive field of two adjacent ON and OFF regions of types 1 and 4. It would respond at light-on and light-off even with a diffuse light stimulus since there is only excitation in either case. A unit with a field of two regions of the types 3 and 6, however, would preferably respond to contrast stimuli, while under diffuse illumination inhibition would cancel excitation.

One reason why types 1 and 3, as well as types 4 and 6, are often not distinguished explicitly may be that in extracellular spike recordings, negative going changes often cannot be observed very easily. In most cases, however, one is interested in opponency rather than the effects of light-on and light-off, and the failure to distinguish them is unfortunate. Obviously, a case of opponency must involve inhibition, and thus at least one region with effects of type 2, 3, 5, or 6. Livingstone and Hubel (1984a) write "To establish the presence of opponency, spatial or chromatic, it is not necessary to evoke ON responses with one stimulus and OFF responses with the other: the opponency is clear if one of the two stimuli evokes a response and both together give a reduced response or none". The reader should not misinterpret the use of the word "necessary" here, or he/she might conclude that the demonstration of ON and OFF responses is sufficient to demonstrate opponency; of course it is neither necessary nor sufficient. The reason why ON-OFF is commonly associated with opponency in receptive fields is probably that the two are correlated in peripheral visual mechanisms, at the level of retinal ganglion and geniculate cells. There, a stimulus that excites at light-on very often inhibits at light-off and vice versa, a principle that has been called reciprocal inhibition. The inhibitory effects are also quite obvious because cells have a relatively high maintained activity. Encoding a signal by impulse frequency in a single channel is naturally asymmetric; by carrying a signal in two opponent channels with reciprocal inhibition the system can transmit positive and negative values symmetrically and can thus achieve a higher degree of linearity than would be possible with only one channel.

At the cortical level this correlation is certainly less strong. In simple receptive fields, several studies have demonstrated inhibitory subregions with no excitatory effect (e.g. Creutzfeldt and Ito 1968; Henry et al. 1969; Bishop et al. 1973; Camarda et al. 1979; Palmer and Davis 1981a; Mullikin et al. 1984a, b; Heggelund 1986b). In two-bar interaction experiments Camarda et al. found that two-thirds of the simple cells had only two excitatory subregions, but 27 of 28 had inhibitory flanks on both sides of the central excitatory region. Heggelund (1986b), using a similar method, also found that nearly all simple cells had at least three subregions. Thus, practically all cells with one and two subregions as judged from overt responses (45% of the simple cells in Heggelund 1986a) must have one or two purely inhibitory subregions. Moreover, across all flanking regions, the correlation between degree of excitation by a single bar and degree of suppression in the two-bar interaction was close to zero (Heggelund 1986b). Reports of excitatory subregions with no inhibitory effect (type 1 or 4) are less frequent (examples in Fig. 4 of Palmer and Davis 1981a and Fig. 2 of Heggelund 1986b). Since the inhibitory effect can only be detected against some activity which must often be evoked by a second "conditioning" stimulus, absence of inhibition in excitatory regions may have gone unnoticed for methodological reasons, because at the position of the conditioning stimulus ON and OFF events cancel physically. The inhibitory end zones of end-stopped cells are other examples of regions producing only inhibition. Examples in the wavelength domain are the "color-opponent center-surround" cells found in the dorsal lateral geniculate nucleus (dLGN) (Wiesel and Hubel 1966) and in layer 4Cβ of the striate cortex (Livingstone and Hubel 1984a; Michael 1985) in monkey in which the center is excitatory for light of one wavelength (either at ON or at OFF), but not inhibitory for the complementary wavelength, while the surround is inhibitory for the latter but not excitatory for the former. Another example is the double opponent cell of Fig. 12 in Livingstone and Hubel (1984a), with a purely inhibitory surround.

Since the combination of ON excitation with OFF inhibition, and its converse, are often found in simple field subregions, the findings of additional inhibitory regions without excitation may appear just as a matter of accuracy and completeness of description. This is how they have generally been interpreted by the authors. As in many physiological studies, the primary interest here was in the mechanism, in this case the way the receptive fields are built up from signals of more elementary units, and less in the functional consequences. One probably gets a better basis for understanding the system by regarding the pairing of excitatory and inhibitory effects as a special effort of engineering made to preserve linearity (Glezer et al. 1982), rather than assuming that it is a general principle with negligible exceptions. For example, a receptive field consisting of two separate ON and OFF regions can only work linearly if, in addition to the excitations caused by light increments and decrements in those regions, increments in the OFF region also counteract the ON excitation and decrements in the ON region counteract the OFF excitation. Thus, the inhibitory receptive field has to be a mirror-image of the excitatory receptive field. One could argue that an inhibitory component could be added at a later stage, by conversion of the excitatory signal of another cell with opposite arrangement of ON and OFF regions. However, this is an assumption that needs evidence. The assumption that purely inhibitory regions are later complemented with excitation is equally questionable. I consider the observations of unpaired regions to be very important, because they reveal essential nonlinearities.

Models of receptive fields illustrate very well the complexity of pattern that can be displayed by a rather simple model network. Heggelund (1981a, b) presented models of simple and complex receptive fields and derived receptive field structures that would be revealed by double-stimulus experiments. Spitzer and Hochstein (1985b) presented a complex field model and derived patterns of grating responses similar to those observed experimentally (Spitzer and Hochstein 1985a). They did not discuss Heggelund's model. The principle feature of Hegeelund's simple cell model is that ON and OFF, excitatory and inhibitory, effects are all derived from just two inputs, one excitatory and one inhibitory, both of which have the same type of concentric center-surround field (either ON or OFF), but are spatially displaced. It is perhaps the most parsimonious simple cell model compatible with those experimental data. In his complex field model, he assumes that either of those inputs comes from both ON- and OFF-center fields with overlapping centers. The complex cell model of Spitzer and Hochstein assumes summation of two or more subunits, each of the center-surround type. Each subunit sums linearly, but their outputs are one-way rectified before the final summation (in the case of two subunits this is similar or identical to either of the input parts of Heggelund's complex field model). Heggelund's models are stationary, whereas Spitzer and Hochstein also modeled the time dependence. While complex receptive fields are nonlinear by definition, the simple field model of Heggelund is also nonlinear, due to the displacement of its excitatory and inhibitory components. Note the similarity with the simple cell mechanism proposed by Marr and Hildreth (1980) for detecting zero-crossing segments. This can be conceived as an "ON AND OFF" circuit, with displaced ON- and OFF-center fields, Heggelund's model as an "ON AND NOT ON" circuit, with two displaced ON center fields. Heggelund (1986a) found that either ON or OFF responses dominated in most simple receptive fields, and that the degree of domination was symmetrically distributed, which is consistent with his model (but not very strong evidence).

It is interesting to compare these theoretical considerations with some recent studies that have provided new direct insight (besides earlier evidence from intracellular recordings, Creutzfeldt and Ito 1968) into the mechanisms of convergence of the ON and OFF systems in the cortex. Schiller (1982), in monkey, and Sherk and Horton (1984), in cat, have studied cortical activity after selective pharmacological destruction of the activity of ON-center retinal ganglion cells. The results were somewhat complex. However, the general observation of reduced responsiveness, which affected not only light-edge but also dark-edge responses, the findings that even some light-edge responses had survived, and that many properties like orientation, direction, and length sensitivity were still approximately normal, are consistent with the idea that ON excitation and OFF inhibition, and the converse, are often paired in order to achieve linearity; one would expect only reductions in dynamic range and linearity in such a system. The finding that large numbers of cells had lost all responsiveness might indicate mechanisms such as the one of Heggelund's simple cell model, in which each single cell receives (directly and via inhibitory interneurons) LGN signals of only one type, ON or OFF. Those cells which received only ON input would then be totally inactivated. That simple cells are not generally connected in this way, or not even as a rule, can be seen also from the studies of Tanaka (1983, 1985), who identified the LGN inputs of cortical cells by simultaneous recordings and cross-correlation analysis of the spike trains. The results clearly showed that ON- as well as OFF-center LGN units may converge with excitatory inputs on a single cortical simple cell (as well as on complex cells).

2.2 "Beyond the Classical Receptive Field"

A number of recent studies have shown that, using more extensive or more complex stimuli, one can often influence the responses of visual neurons in a stimulus-specific way from a much greater retinal area than expected from the analysis with simple, localized stimuli. Common to these studies is that visual context information appears to be incorporated in the responses of a neuron, a feature that often appears as an addition to (or modulation of) local receptive field properties. Allman et al. (1985b) have recently reviewed the literature under this aspect, considering, among others, the phenomena of inhibitory and facilitatory effects produced by gratings or random patterns in neurons of cat area 17 (e.g., Maffei and Fiorentini 1976; Nelson and Frost 1978; Hammond and MacKay 1981; Nelson and Frost 1985), responses to anomalous contour stimuli in rhesus monkey V2 (von der Heydt et al. 1984; see below), color-coding properties (Zeki 1983a, b; see below) and other surround effects in rhesus monkey V4 (Desimone et al. 1985), as well as foreground-background interactions between moving patterns in the owl monkey middle temporal area (MT) (Allman et al. 1985a).

While the areas whose stimulation produced these effects belong to the receptive field by definition, several studies have defined a "classical receptive field" (or similar) and have taken pains to demonstrate that these effects originate outside this field. The definition of what is classical varies, however. Maffei and Fiorentini (1976) simply define it as the region within which a moving or flashing bar can elicit a response from a cell, and consequently speak of the rest as the "unresponsive regions". What they show is that stimuli (mostly gratings covering surrounding regions) that are ineffective by themselves, can nevertheless markedly influence responses when combined with other stimuli. Allman et al. (1985a) also use this definition and point out that it is wider than that of the "excitatory receptive field" because conventional local stimuli may also inhibit the activity below the spontaneous firing rate, for example when moving against the preferred direction. Nelson and Frost (1978, 1985) use a more extensive interpretation, including not only the regions in which a moving bar by itself can influence the cell's activity, but also those were it can modify the activity evoked by a second, "conditioning" stimulus. They thus wanted to include as classical the inhibitory parts of the fields. Inhibitory subregions are, of course, part of the simple receptive fields and inhibitory end-zones of the hypercomplex fields of Hubel and Wiesel (1962, 1965). Thus, while the definition of Maffei and Fiorentini is operationally simple, it does not include all that has become classical. The other definition does include all that, but is extremely difficult to verify (see Nelson and Frost 1978) and may also involve logical problems. I think one could avoid some of the classical confusion simply by abandoning the term "classical receptive field". It is a misconception anyway, because what is classical, after all, is a concept of receptive field.

Since anomalous contours often appear at sites where the stimulus is actually empty, the assumption is common that neural responses related to such perceptions should be evoked by stimuli entirely outside the classical receptive field (whatever this may include). This may lead one to expect that a neuron should respond to stimuli confined to unresponsive regions, a self-contradictory postulate. I think such an assumption is misleading and unnecessary. It probably arises from the idea of the "cortical image", where the neural activity across the cortex is thought to represent an image which is then somehow analyzed. The responses to a stimulus of two abutting gratings (see Fig. 6) reveal mechanisms beyond classical receptive-field concepts even though the stimulus covers the cell's response field. This kind of contour is as anomalous as the one of Fig. 7 or the contours in the famous Kanizsa triangle, and they probably all involve the same mechanism. What von der Heydt et al. (1984) have shown in this case, and what Zeki (1983a, b) and Allman et al. (1985a) have shown in others, is that a neuron's responses may depend strongly on the image context, thus being influenced by stimulation from large areas, but may still appear to carry a "position tag", i.e., they seem to signal a local property, such as site and orientation of object boundary, or surface color, or movement. This interpretation may be based on observations that the response-evoking stimuli always produce a certain perception in human observers, such as color or contour, at some point in the visual field. For this, the existence of a classical receptive field may not even be necessary (see also the discussion in Zeki 1983a, of color-coded responses that cannot be evoked by isolated chromatic stimuli). To find out what these neural signals could mean for a subsequent processing stage is, of course, one of the major difficulties of this research.

2.3 Some Questions About Receptive Field Types and Columns

I should not conclude the discussion of receptive-field concepts without mentioning a few questions on which the evidence seems to be contradictory.

Is orientation constant in *columns*? Recording from V1 in the alert monkey, Bauer et al. (1983) found that the cells' preferred orientations frequently shift at the border between layers 4 and 5, in penetrations normal to cortical surface. Shifts of 45°–90° were found in 70% of the penetrations. Krüger and Bach (1982) made a similar observation. E. Peterhans and I have seen such shifts too (unpublished). Similar shifts have been found in cat (Bauer 1982). Other studies do not report these shifts.

Are there few or many *unoriented cells* in monkey striate cortex? Once it was thought that only cells in layer 4C had circularly symmetric receptive fields (Hubel and Wiesel 1968, 1974). Although 4C is a layer densely packed with cells, this could only be some smaller fraction of striate cells. (It is now held that only the lower part of this layer contains concentric cells, while those in the upper part are oriented, simple cells; Blasdel and Fitzpatrick 1984; Hawken and Parker 1984; Livingstone and Hubel 1984a; Michael 1985.) Recently, Livingstone and Hubel (1984a) have found many cells with unoriented fields also in other layers, especially 2 and 3, in the cortical patches called blobs. In their Table 1, which lists cells recorded in layers 2 and 3, 36% of the cells of the foveal representation (2°–2 1/2°) and 20% of the nonfoveal cells were unoriented. Zeki (1983c), sampling from all layers, finds that over

50% of cells in the central 0°–5° are not orientation selective, a proportion that decreases with eccentricity to 20% at 10°–30°. If wavelength-selective cells (which were typically unoriented according to these studies) are excluded, there still remain 26.2% in the central 0°–5° which were neither orientation nor wavelength selective. Zeki cites Poggio et al. (1975) who also reported over 50% non-orientation-selective cells in the central projection area. However, Poggio et al. (1977) found only 15% unoriented cells [43 of 240 that responded to their stimuli (gratings), table 1, in Poggio et al. (1977)]. One obvious difference between these studies is that Poggio et al. (1977) recorded from alert, active monkeys, the others from paralyzed monkeys. Poggio et al. (1977) state also that unoriented cells were found in all layers, and only relatively more frequently (but still in a minority) in layer 4C. Other authors also find unoriented cells, but obviously less frequently than Zeki. In Fig. 8 of De Valois et al. (1982), I counted 13 unoriented cells (orientation bandwidth > 180°) from 168, i.e., only 7.7%. Even of the "color-coded" cells the great majority was found to be oriented (Michael 1978a, b, c, 1979); of 863 classified cells only 137 (16%) were concentric. Thorell et al. (1984) mention eight concentric among 42 "simple" cells (by the modulation criterion, see below) which in turn made up slightly more than half of their sample. Thus, again, only about one-tenth of the cells were concentric. Livingstone and Hubel (1984a) used *Macaca fascicularis,* while Michael (1978a, b, c, 1979) used *Macaca mulatta,* but Thorell et al. (1984) and Michael (1985) also used *Macaca fascicularis.* Besides the differences in classification (terms like "unoriented", "concentric", "unsensitive", and "unselective for orientation" may have different meanings), a more severe disagreement is apparent. Is the ever-invoked sampling bias of the microelectrode to be blamed again? Are differences in use of narcotics important? From my own experience in alert animal experiments I favor the view that unoriented cells are few.

Are there few or many *simple cells* in monkey striate cortex? While estimates of the proportion of simple cells in cat striate cortex ultimately seem to converge (55%, excluding a small fraction of nonoriented cells, in Palmer and Davis 1981a; 61% in Dean and Tolhurst 1983; 63%, when B cells are included in the complex group, in Camarda et al. 1985a; 69% in Heggelund 1986a), there is vast disagreement between the monkey studies. For example, De Valois et al. (1982) found simple and complex cells in a ratio of 61:39, while Hubel and Wiesel (1968) found only 9% simple and 65% complex, or a ratio of 12:88. De Valois et al. point out that their figure is virtually the same as the proportion found in cat, and quote Schiller et al. (1976a) for a simple to complex ratio of 42:58

in the monkey. However, Poggio (1972), Dow (1974) and Gouras (1974) also found few simple cells. Foster et al. (1985) found 14% simple and quote Schiller et al. (1976a) with a proportion of 22%. As a matter of fact, Schiller et al. (1976a) classified as simple 245 cells (22%) of their total sample of 1125 which included 297 unclassified, 130 unresponsive and 111 unoriented or poorly oriented cells, and emphasized that many cells were not classified simply because the data were incomplete in this respect. Thus, only the simple/complex ratio, as quoted by De Valois et al., may (perhaps) be compared with the figures of these recent studies (which do not give numbers of unoriented cells).

Unlike previous studies, De Valois et al. (1982) have used a strictly quantitative criterion for *classification,* the modulation of responses to drifting gratings (first harmonic component greater than mean impulse rate). They found "while there are cells which are difficult to classify, we found that every cell classified as simple cell by Hubel and Wiesel's criteria was classified as X-cell by Enroth-Cugell and Robson's criteria" [i.e., the classification by grating responses, either stationary (null position) or drifting (modulation)]. Consequently, they equate the classes of X and Y with simple and complex. The above statement obviously does not mean that Hubel and Wiesel's method of mapping receptive fields with bars would always lead to the same result as the grating test, since the "difficult" cells might be classified differently. The unoriented cells, for example, were apparently included in the simple/complex groups by De Valois et al. (probably around 10%, see above), but not by the previous studies. Could it be that complex cells are often classified wrongly by the modulation criterion? Comparing various criteria, such as discreteness of receptive field regions, spatial summation, and modulation of grating response, in *cat,* Dean and Tolhurst (1983) conclude that none of these criteria alone allows complete unequivocal classification, pointing out that many simple cells had intermediate modulation indexes. However, their results also indicate that a classification by response modulation would correlate highly with the field mapping results, indeed better than the two other predictors examined. From Fig. 5 of their study, taking a modulation index greater than 0.8 as the criterion for simple (as suggested in the discussion), I calculate that only 41 of 381 cells (11%) would change classes compared to the original classification based on mapping; the agreement was even better for the subsample of cells that were mapped quantitatively. Moreover, the wrongly classified cells were mostly simple, i.e., the modulation criterion would result in underestimation of the proportion of simple in this case. Some ambiguity could arise from the fact that the modulation of complex cell responses depends on spatial frequency (Spitzer and Hochstein 1985a). In the *monkey*, Foster et al. (1985) have used techniques similar to those of De Valois et al. (1982) and also found that responses of all conventionally classified simple cells were modulated completely (14%), while those of complex cells were unmodulated (72%) or partially modulated (12%). However, distributions of a modulation index were not presented, and classification was not a major concern of this study. Differences between foveal and parafoveal samples were minor in both studies. Poggio et al. (1977) classified cells in the active monkey on the basis of square-wave grating responses into modulated and fusional (which is obviously related to the classification of De Valois et al.) and found a ratio of 34:66, very similar to the figure of Schiller et al. (1976a) and not too far from that of De Valois et al.

Some grating studies have to be interpreted with caution here because they seem to have used gratings of fixed length (Dean and Tolhurst 1983; Foster et al. 1985; Spitzer and Hochstein 1985a), whereas De Valois et al. (1982) apparently adjusted the length of the grating. Poggio et al. (1977) also used fixed length gratings, but state the numbers of unresponsive, unclassified, and unoriented cells. It is not clear if end-stopped cells were entirely or partially excluded in those other studies. In any case, this detail, which is not all obvious from the papers, may have affected the results significantly since end inhibition is a frequent phenomenon.

Thus, the disagreement is hard to explain. Clearly, if simple cells in monkey were as few as some studies suggest, this would mean that they are much less important in primates than in cat, and studying simple cells in cat would not tell us as much as we had hoped about human vision.

This classification, of course, has a long history (see Kato et al. 1978 for a review of older cat studies). Roughly speaking, in the initial studies of Hubel and Wiesel (1962, 1965, 1968) receptive fields were analyzed qualitatively, mainly with stationary flashing stimuli. When quantitative methods using averaged responses were introduced (perhaps starting with Pettigrew et al. 1968), they were limited for some time to moving stimuli. Movement, of course, complicates the situation because time dependence may confuse the spatial analysis. Apart from sporadic earlier attempts, extensive quantitative spatial analysis with stationary stimuli has been undertaken only more recently (e.g., Emerson and Gerstein 1977a, b; Movshon et al. 1978a, b; Camarda et al. 1979; Kulikowski et al. 1981; Palmer and Davis 1981a, b; Heggelund 1981a, b; Mullikin et al. 1984a, b; Camarda et al. 1985a, b; Heggelund 1986a, b).

Some authors (see Bishop 1984) have emphasized that a fraction of cells (some 15%) in cat striate cortex do not respond at all, or only weakly, to stationary flashing stimuli but respond strongly to moving stimuli. Other studies using stationary stimuli do not mention this limitation (Palmer and Davis 1981a; Heggelund 1986a). Regarding Hubel and Wiesel's criteria for simple cells, some studies have found that a single parameter specifying discreteness of ON and OFF zones or spatial summation along the width dimension did not allow unequivocal classification of cells in cat striate cortex (Dean and Tolhurst 1983; Tolhurst and Dean 1985; cf. also Hagerty et al. 1982). However, Heggelund (1986a), using averaged flashing bar responses, did find two nonoverlapping distributions in the separation of ON and OFF zones (for all cells that had at least two zones, i.e., 315 cells out of 322 studied).

The category of simple cells can be subdivided according to the pattern of subregions, contrast preference, directional selectivity with moving stimuli, and interaction between contrast and direction of movement (Schiller et al. 1976a) or simply according to the number of excitatory regions (Palmer and Davis 1981a). It is confusing that Palmer and Davis use the same names (S1 .... S4) as Schiller et al. with different definitions without a *caveat.* While the classification of Schiller et al. looked promising for a functional interpretation, that of Palmer and Davis is, of course, easier to remember.

For a recent review of attempts to split the categories of simple and complex further see Bishop (1984). The history of B cells was also reviewed in Kulikowski and Bishop (1982).

Alert Versus Anesthetized. As in the above discussion of orientation selectivity, one sometimes has the impression that there are important differences between experiments with anesthetized-paralyzed animals and alert animals. Many reports emphasize that "interlaced" presentations of various stimulus conditions are necessary to compensate spontaneous fluctuations in responsiveness. In my experience, recordings in the alert monkey are not plagued by such unexplained fluctuations. Recently, I read that in a study of the former type in cat, artificial respiration with 100% oxygen helped to maintain the proportion of end-inhibited cells, which otherwise tended to decline during the experiment (Bolz and Gilbert 1986). The possibility of such changes is largely excluded in chronical alert animal experiments in which the *Dura mater* is left intact except for penetrations by the microelectrodes. In the alert monkey, von der Heydt et al. (1986a) found cells in the upper layers of V1 responding selectively to periodic patterns, and not to bars and edges (see below). Such cells have not been described in anesthetized animal studies.

2.4 Plasticity of Receptive Fields

That problems in vision are often empirical questions about the statistical structure of the visual world was perhaps realized first by Helmholtz and has become a basic notion in the field of computer vision. "Studying vision is a mixture of studying processes and studying the world from this rather specialized point of view – something that natural evolution has been doing for a long time" (Marr 1982). This is one reason why cortical receptive fields are so interesting. They can be regarded as the result of a long process of optimization. The question has often been asked as to whether cortical receptive fields are in fact entirely a product of evolution, or whether experience of the individual organism also plays a role.

As can be seen from the review by Movshon and Van Sluyters (1981), the argument about whether visual experience plays an instructional role in the development of the visual cortex or is merely a necessary condition for the formation of an innate pattern of connectivity has not really been settled. An example strengthening the latter point of view is the study by Carlson et al. (1986). In young monkeys one eye was closed shortly after birth, while the other was allowed vision of lines of mainly one orientation. As a result, cortical cells dominated by the open eye showed the normal, full range of orientations, while among cells dominated by the closed eye the orientations by which the open eye had been stimulated were missing. Thus, similarly in principle to the study in kittens by Rauschecker and Singer (1981), this experiment simultaneously demonstrates that monocular occlusion causes deprivation (for the orientations that were present in the environment), and that the pattern of environment is otherwise irrelevant (for the cells driven by the open eye).

Movshon and Van Sluyters (1981) took pains to defend the view of an active role of the environment by arguments from many studies, none of which was overwelming. But they omitted discussion of a paper that claimed to provide unambiguous evidence (Shinkman and Bruce 1977). In this study, it was shown that visual experience through goggles that rotated the images in the two eyes relatively to each other by a small angle, changed the orientation preferences in the two eyes of binocular cortical cells of kittens accordingly. When I first read this paper, I thought that this result was not really evidence for a plastic change in the visual cortex, since the relative alignment of the two eyes could have been modified. How could one know the exact angle the kittens would have developed without disparate experience? Similar criticism had been raised against Shlaer's (1971) demonstration of plasticity in binocular correspondence of receptive field positions. The apparently normal mean angle between the slit pupils in a number of kittens (Bruce et al. 1981) may have removed the doubt to some extent. Hänny and von der Heydt (1982) and Dürsteler and von der Heydt (1983), again in kittens, have studied the effects of disparate image deformations, which could not be cancelled by any rotation of the eyes. In the former study, the disparity resulted from an unusual geometry of the environment, in the latter, from image deformations by pairs of Ames lenses. Corresponding binocular receptive field changes were found in both cases. In the latter study, it turned out that orientations as well as positions of receptive fields had been modified so as to compensate the imposed disparities quantitatively. Here, then, are experiments showing that development takes account of detailed information about the environment, namely the complete pattern of disparity in the visual field. The observation of a tilt of the vertical horopter (Helmholtz 1864; Cogan 1979) indicates that similar processes may govern the development of human vision (von der Heydt et al. 1980). But, does plasticity really improve vision? Perhaps the fly will tell us (Mimura 1986).

3 Spatial Filtering

Perhaps the most stimulating idea about visual cortical function has been the Fourier analyzer hypothesis. Following the success of the application of Fourier analytical methods in psychophysics (Campbell and Robson 1968), it was proposed that the visual cortex begins visual processing by transforming visual space into a frequency representation, and henceforth handles visual information in terms of the frequency components (Pollen et al. 1971; Glezer et al. 1973; Maffei and Fiorentini 1973). Recent reviews of this theory and its physiological evidence can be found in De Valois and De Valois (1980), Robson (1980), Pollen and Ronner (1983), and Shapley and Lennie (1985). For the psychophysics see Kelly and Burbeck (1984) and Westheimer (1984).

3.1 Selectivity

The most obvious requirement of the Fourier analyzer theory is the neurons' selectivity for the spatial frequency of gratings. In monkey striate cortex Schiller et al. (1976b) determined frequency response functions, and De Valois et al. (1982) have now measured contrast sensitivity functions, i.e., the grating contrast required for a criterion response at each spatial frequency which can be compared better with psychophysical measurements. Peak frequencies and widths of these functions were distributed similarly for simple and complex cells (definition by response modulation, see above) and for foveal and parafoveal samples. In the latter, the distribution of peaks was shifted only slightly to lower frequencies compared to its considerable spread, indicating that a wide range of preferred frequencies is represented at each eccentricity. Median bandwidths of 1.4 and 1.5 octaves were found in simple and complex cells respectively, but the enormous variations in bandwidths (from about 0.5 to over 2.5 octaves) should make one cautious about speaking of the typical bandwidth of cortical neurons. Correlations were found between widths and peak frequencies (cells with high preferred frequency tend to have smaller bandwidths in octaves) and between the widths of frequency and orientation tunings (sharply oriented cells tend to be frequency selective, cf. Tolhurst and Thompson 1981 for the cat). Based on the tuning curves for spatial frequency and orientation, cells were characterized by areas in the two-dimensional frequency plane (see discussion below).

Foster et al. (1985) have confirmed these results on spatial frequency tuning in monkey V1, extended them by recording temporal frequency tunings as well, and have also obtained comparable data from V2. Comparing V1 and V2, they find similar widths of tuning but lower preferred frequencies in V2 in the spatial domain. In the temporal domain the tuning properties were different in that cells in V1 showed mostly low-pass characteristics or were broadly tuned, while a majority of neurons in V2 showed bandpass tuning (mean bandwidth of 2.1 octaves for temporal frequency). Combinations of spatial and temporal frequencies were varied and little interaction was found. The temporal tuning properties found in V2 were discussed and related to psychophysics within the framework of spatio-temporal filtering. Note, however, the limits to generalization posed by nonlinearity in temporal summation (Tolhurst et al. 1980). The results on spatial preferences in V1 and V2 were found to be similar to the 17–18 differences found in cat (Movshon et al. 1978c; Berardi et al. 1982). Regarding the results on spatial frequency selectivity, it was argued that for certain types of information no further elaboration occurs from V1 to V2. The shifted ranges of preferred spatial frequencies between the two areas were interpreted as providing an extended total range, in accordance with the suggestions of a hierarchy of filtered images at different scales (Campbell and Robson 1968; Ginsburg 1975, 1982). I doubt that this concept of V2 function will be of much help, and in general, that V2 functions can be adequately described by spatial and temporal filter characteristics at all (see also Contour).

Since most cortical cells are selective for *orientation and spatial frequency* one can ask whether the tuning of a cell to one of these parameters depends on the choice of the other. Most frequently, spatial frequency tuning curves have been recorded at some previously determined optimum orientation, and likewise, orientation tuning curves have been recorded using gratings of some previously determined optimum frequency or using other, aperiodic, stimuli. Thus the question arises whether these two measurements depend on each other, or whether receptive fields can actually be described by the separately determined orientation and spatial frequency tuning curves (Daugman 1980, 1984).

Independence of the two (or "polar separability") had been demonstrated before for a few examples, but the problem was thoroughly examined first by Webster and De Valois (1985), who recorded the responses of each single unit at various combinations of frequency and orientation. Their results show that while the optimum orientation of a cell is generally the same at all spatial frequencies, the optimum frequency usually depends somewhat on the orientation: at orientations on either side of the optimum, the peak of the frequency tuning was found displaced to lower frequencies. The amount of displacement depended on where a cell was more selective; pronounced shifts were found only in cells relatively more selective for orientation than for spatial frequency. The mostly small deviations from separability and their dependence on the relative selectivities of a unit were interpreted to favor a two-dimensional Gabor function model rather than the aligned LGN fields model of Hubel and Wiesel (1962). Where model predictions differed from neuron data, both models seemed to predict greater frequency shifts than actually observed.

3.2 The Question of Linearity

The spatial frequency analyzer theory requires not only that cortical neurons are selective for spatial frequency, but also that they function, at least approximately, as linear filters. The essential property of such a filter is that its response to any stimulus that can be conceived as a sum (optical superposition) of other stimuli equals the sum of the responses to the individual components. Its responses can thus be predicted from its responses to other stimuli, for example those to sine wave gratings. In fact, sinusoidal gratings are special only because the response of a linear filter to such a grating is again a sinusoidal grating. The response can thus be described simply by attenuation (or gain) and phase shift, i.e., a pair of numbers for each frequency (the complex transfer function). In other words, sinusoidal gratings are the eigenfunctions of linear filters. How can neuronal responses be conceived of as a grating? The responses to a sinusoidal stimulus presented at various positions (phases) should vary as a sinusoidal function of the phase angle.

If the assumption of linearity is not given, sine wave tuning characteristics have only descriptive value, characterizing the responses to gratings of that particular profile and contrast. The question of linearity is thus of more general interest than the spatial frequency analyzer theory.

3.3 Prediction of Receptive Field Profiles

It was of particular interest to see if the cells' receptive field profiles or "line-weighting functions" could be predicted from their sine wave grating responses. Several studies have compared such a prediction with the measured profile (De Valois et al. 1978; Movshon et al. 1978a, b; Maffei et al. 1979; Kulikowski and Bishop 1981b; Glezer et al. 1982) or with the qualitative receptive field map obtained by listening to the responses to stationary flashing bars (Andrews and Pollen 1979). The general conclusion was that the profiles of most simple cell receptive fields can be well-predicted from their spatial frequency tuning by means of the inverse Fourier transform as though they were linear spatial filters. Before describing the conclusions in more detail, I shall review some critical methological points.

The validity of such a comparison may be limited for a number of reasons (cf. Movshon et al. 1978a). Contrast sensitivity has often been compared with response strength (Movshon et al. 1978a, b; De Valois et al. 1978; Kulikowski and Bishop 1981b). Indeed, Movshon et al. (1978c) found in a population of cells that measurements of tuning from response and sensitivity functions correlated fairly well. However, the two functions are not identical in every cell. Furthermore, the different units of measurement imply arbitrary normalization. Moving gratings have been compared with stationary flashing bars (Movshon et al. 1978a,b; De Valois et al. 1978; Andrews and Pollen 1979). Comparing moving and stationary stimuli is awkward, particularly since most simple cells are sensitive to the direction of motion. Often, constant temporal frequency, and thus varied drift velocity, was chosen for the gratings, while the Fourier components of the bar, of course, all moved at the same velocity. Only two studies compared gratings moving at constant velocity with moving bars (Maffei et al. 1979; Gletzer et al. 1982). However, a model without time dependence is clearly inadequate (Movshon et al. 1978a; Tolhurst et al. 1980). The phase of grating responses was determined only in one study (Glezer et al. 1982), all others used only amplitudes (response strengths or sensitivities) for the Fourier transform, choosing arbitrary (often frequency-dependent) phase behavior so as to achieve the best fit. Thus, looking for a strict comparison between receptive field profile and back-transformed spatial frequency tuning, one is left with a few examples in a single study, and these even show moderate (but apparently significant) differences rather than agreement.

I find the scarcity of Fourier phase measurements particularly hard to understand since the phase of responses to drifting gratings is readily available, at no cost in experimental time. Perhaps a lack of confidence in the stability of eye position prevented experimenters from attempting phase measurements; on the other hand, it is usually assumed that subregions of receptive fields, and their relative positions and overlaps, can be determined exactly under these conditions. Studies including phase measurements are to be expected (Hamilton et al. 1985).

And why has a comparison between measured and predicted profiles never been carried out on the basis of responses to stationary bars and stationary gratings? The use of stationary gratings in sine and cosine phase would allow determination of spatial phase as well. Stationary modulated gratings have been used to disentangle temporal and spatial properties (Movshon et al. 1978a; Spitzer and Hochstein 1985a). Especially for any modeling of cortical functions it would be important to know which of the various simple cell types can in fact be regarded as linear, and of those that cannot, what is the type of nonlinearity.

I should also point out the peculiar use of the term "linear" in connection with receptive fields. Receptive fields are usually called linear when responses are compatible with a model of linear spatial summation and output nonlinearity. Since impulse frequency is used as a measure of response strength, and since the maintained activity of most cortical units is zero or a low frequency, a rectification nonlinearity is obvious. For this reason, responses (increments in spike frequency) to stimuli of opposite contrast polarities (light and dark bars, or mirror-symmetric edges, or the onset and offset of one of these) are usually plotted with opposite sign in order to obtain the profile (or weighting function) of a receptive field; e.g., ON responses are plotted upward on the ordinate, OFF responses downward. Thus, negative parts of the function do not indicate a subtractive effect (inhibition), but rather a positive effect (excitation) produced by a complementary stimulus. Thus, the assumption is made that the "neurone's responses at any position could be interpreted as showing inversion when stimulus polarity was inverted, neglecting an obvious rectification" (Movshon et al. 1978a, p 70). Often, this assumption is implicit. It is another example of the practice already discussed obove, namely of not distinguishing between ON inhibition and OFF excitation, nor between OFF inhibition and ON excitation. There are of course cases where a cell gives incremental responses to both light and dark bars. This is typical for complex cells, but also for a minority of simple cells, described as "non-linear simple cells" (Movshon et al. 1978a). Many simple cells respond to both polarities at some positions, thus showing some overlap of ON and OFF regions (Maffei et al. 1979; Kulikowski and Bishop 1981b; Palmer and Davis 1981a; Glezer et al. 1982; Dean and Tolhurst 1983; Camarda et al. 1985a; Heggelund 1986a). In cases of such double responses the function R(s)-R(-s) is regarded as the line-weighting function, where s denotes the stimulus and R(s) the response.

One may add that a simple cell model that neglects rectification non-linearity is useful only if one assumes that these signals always come in antagonistic channels, and that only R(s)-R(-s) is used at subsequent processing stages (i.e., signals from two channels should always act in pairs of excitatory and inhibitory inputs, like complementary transistors in a push-pull amplifier). Incidentally, there seems to be no evidence for this; on the contrary, if complex cells receive any input from simple cells, as Hubel and Wiesel (1962) thought, this is apparently not of the above, symmetrical form, otherwise their receptive fields would not be so different.

Another "trick" frequently used by physiologists when recording receptive field profiles is to apply a "conditioning stimulus" which serves to produce an artificial "background discharge" against which both excitatory and inhibitory effects of the stimulus can be observed (Henry et al. 1969). The conditioning stimulus may be a bar oscillating over the most responsive part of the receptive field, or a visual dynamic noise (Maffei et al. 1979). This stimulus is not time-locked to the test stimulus and therefore its responses appear as an elevation of the baseline in the averaged response to the test stimulus. The absence of spontaneous activity (especially in simple cells) is usually the reason given for applying this technique. It helps, of course, to conceal nonlinearities, but it allows us to see, qualitatively, inhibitory effects of the test stimulus.

Movshon et al. (1978a) found good agreement between predicted and measured line-weighting functions. In particular, they point out that all the predicted "ripples" of the function corresponded to observed ON and OFF regions in every case in which the measurements of the line-weighting function extended far enough. They found agreement even in nonlinear simple cells, i.e., those in which presentation of a stationary grating elicited responses over more than 180° of phase positions, which constituted about one-third of the simple cells tested. The line-weighting functions of these cells were mostly double valued, since they responded to light and dark bars at the same positions. But the difference between the two followed the curve predicted from the spatial frequency sensitivity function. The authors conclude that "under certain conditions, a linear model of the simple cell's receptive field is statisfactory". Clearly, however, this statement was meant to stress the difference between these results in simple cells and those obtained in complex cells (Movshon et al. 1978b). Stressing another aspect of their results, one could also say that sinusoidal grating stimuli often bring out linear behavior of otherwise nonlinear receptive fields.

Maffei et al. (1979) demonstrated an example of excellent agreement between measured and predicted profiles. However, they also qualified this result by saying that their theoretical and experimental data were only qualitatively similar, and that the example shown was the one with the best fit out of ten. Nevertheless, they believe that their results "suggest that visual cortical neurones behave, at first approximation, as a linear system". This is a paper with many subtleties, some of which will be discussed below.

Glezer et al. (1980, 1982) have measured phase and amplitude of the grating responses, and presented examples of simple cell receptive fields in which measured and predicted profiles were clearly different. The most obvious deviations consisted in over-estimations in the prediction of the negative portions of the profile. The predicted function could even show a large negative lobe where almost no dark bar responses were obtained (Figs. 1 and 2 of Glezer et al. 1982).

Pointing in the same direction, Andrews and Pollen (1979) found that spatial frequency tuning (moving gratings) often led to more side lobes of the predicted profile than mapping with stationary bars had indicated.

In order to determine the true extent of spatial summation, De Valois et al. (1985) have therefore measured responses of cortical cells to sinusoidal gratings of various numbers of cycles or half-cycles, embedded in a homogeneous field of equal mean luminance. In most cells the strongest response was obtained at some optimum number of cycles between one and seven, with an average of 2.5–3 cycles. (Unfortunately, these authors seem to have a strange difficulty in distinguishing between cats and monkeys; about one-third of the cells are said to come from monkeys, and the rest from cats.) Thus, the optimum number of half-cycles was five or six on average, whereas simple receptive fields with that many excitatory subregions are rare, a maximum of six has been reported (Mullikin et al. 1984b).

The number of subregions is theoretically related to the width of spatial frequency tuning; sharper tuning requires more lobes in the receptive field weighting function. In the case of linear filters with Gabor function profile (i.e., sine or cosine functions with Gaussian envelope) the width of frequency tuning is inversely proportional to the extent of spatial summation (cf. Marcelja 1980, who noticed the similarity between these functions and simple cell receptive field profiles; a table of the expected numbers of side lobes for a given tuning can be found in Kulikowski et al. 1982).

Andrews and Pollen (1979) have shown how the frequency tuning changes when the profiles obtained from the frequency tuning are set to zero outside the responsive part of the receptive field (the part that could be mapped) and then back-transformed. In many cases this modified tuning was markedly different from the measured tuning, having smaller peak amplitude and greater width, and a shallower low-frequency cut-off. The sharper tuning was thus attributed to unresponsive regions of the receptive field.

De Valois et al. (1985) considered the existence of subliminal side lobes to be the cause of a general disagreement between spatial frequency tuning and extent of mapped receptive fields in cortical neurons. The simple receptive fields with two or three subregions described by Hubel and Wiesel (1962), it was argued, could only have bandwidths of 1.6 octaves and more, whereas fewer than half of striate cortical cells actually have bandwidths this broad. Indeed, Movshon et al. (1978c), in cat, and De Valois et al. (1982), in monkey, found means or medians of about 1.5 octaves for simple cells and only slightly more for complex cells; many cells had bandwidths of less than one octave. De Valois et al. (1985) argued that the extended summation found in their experiment indicates additional, subliminal regions which would account for the sharper tuning and suggest a mechanism of linear spatial summation with a high output threshold. Maske et al. (1985, p. 683 ff.) also state that receptive field profiles obtained with stationary bars fell short of the true extent of simple fields, while measurements with moving bars (Kulikowski and Bishop 1981a) showed better agreement and suggest the same explanation. How such a mechanism could be reconciled with the observations of high contrast sensitivity (contrast thresholds below 0.2% in Fig. 4 of De Valois et al. 1982; for contrast-response functions see Dean 1981 and Albrecht and Hamilton 1982) remains unclear. De Valois et al. (1985) point out that intracortical inhibition and/or interactions between spatial frequency channels (De Valois and Tootell 1983) could be responsible, in part, for narrowed frequency tuning and hence for extended grating summation; truncated gratings have a distributed power spectrum, and extending a grating concentrates its power near the grating frequency. As a result, cross-frequency inhibition should be diminished.

The discrepancies between the predicted and observed numbers of lobes of receptive field profiles and the general finding of sharper frequency tuning than expected from the number of excitatory subregions obviously parallel the findings discussed above, that many receptive fields had inhibitory subregions that only appear as interactions in two-bar experiments. Indeed, Glezer et al. (1980, 1982) considered failure in pairing of inhibitory and excitatory subregions, or spatial displacement between such pairs, as the major causes of nonlinearity in simple cells (besides rectification). Cases were inhibitory regions do not produce excitation (unresponsive regions) are most obvious, but quantitative differences, e.g., weaker excitation than inhibition, may also be important. This explanation is different from that of subliminal subregions, which assumes a high threshold or "negative maintained discharge level" (De Valois et al. 1985). The unpaired inhibitory regions can be activated by low-contrast stimuli but do not constantly produce inhibition. As a consequence, this explanation is not in conflict with the finding of low contrast thresholds. Although the afore-mentioned reciprocal relationship between tuning width and extent of spatial summation was derived for linear filters, it is obvious that the addition of unresponsive inhibitory regions also results in narrower spatial frequency tuning than predicted from only the responsive regions (see Andrews and Pollen 1979). For this effect the unresponsive regions must be properly related in width and phase to the other regions of the receptive field. The interaction plots of Heggelund (1986b) clearly show such a relationship. The finding of De Valois et al. (1985) that, in many cells, increasing the number of cycles of a grating beyond the optimum decreased the responses may also be related to those unresponsive inhibitory regions (although other, phase-independent mechanisms might also be involved).

Besides the existence of purely inhibitory regions one also has to consider as sources of nonlinearity regions where excitation does not match inhibition quantitatively. ON inhibition, for example, may be stronger than OFF excitation at the same location and may also be evoked from a wider region. Of the inhibitory flanks on either side of the dominant excitatory region, revealed by two-bar interaction, the stronger one was wider on average than the dominant region; indeed, it was often more than twice as wide (Heggelund 1986b). One may assume that it was also wider than the corresponding (nondominant) excitatory region, since nondominant excitatory regions were smaller on average than the dominant excitatory regions, but a direct comparison was not made (often there was no excitation to compare with). "Dominant" means the type (ON or OFF) of the strongest response of a cell, and "nondominant", the opposite type of response.

This tendency was also obvious in earlier studies (Bishop et al. 1973; Camarda et al. 1979). For example, Camarda et al. found that the overall width of simple receptive fields including inhibitory side bands was about three times larger than the overall width of only the excitatory regions; Heggelund found a factor 2.3. Considering this evidence, the now fashionable description of simple receptive fields by Gabor functions appears as a backward step in understanding, since, besides linearity (matching of excitation and inhibition), it implies constant widths of the subregions (cf. discussion in Webster and De Valois 1985).

De Valois et al. (1985) also found the negative correlation expected from the linear theory mentioned above between number of subregions and spatial frequency bandwidth in simple cells: broadly tuned cells never preferred more than two cycles and sharply tuned cells seldom peaked at fewer than two cycles. By contrast, complex cells were sometimes broadly tuned but responded best to gratings of several (up to seven) cycles. The authors suggested that these cells sum the signals of several broadly tuned simple cells with receptive fields in neighboring loci, and without regard to spatial phase (cf. also Movshon et al. 1978b; and the model of Spitzer and Hochstein 1985b).

It seems to me that the negative correlation in the receptive field data of De Valois et al. ($r = -0.66$ for simple cells) is not as strong as one would expect from the simple inverse relationship between frequency band width and spatial width in Gabor functions. For example, cells with optimum widths of two cycles had bandwidths between 0.8 and 2.1 octaves. This is again difficult to explain in the summation-to-threshold concept, but follows naturally from the assumption of purely inhibitory regions. The threshold would not affect the estimate of spatial summation (the idea of the experiment carried out by De Valois et al.), and variations in threshold would have only a minor influence on the estimates of tuning width of the underlying weighting function. The addition of inhibitory regions to a receptive field, however, would sharpen the tuning without changing the peak of the summation curve.

As already mentioned, predictions of profiles from receptive field analysis with sinusoidal gratings have also been calculated for complex cells (Movshon et al. 1978b; Glezer et al. 1980). As might be expected, the predicted profiles were very different from the measured bar response profiles, the latter being much broader. Movshon et al. found that the predicted profile corresponded well to interaction profiles obtained with two flashing bars. These profiles were interpreted as characterizing subunits of the receptive field.

3.4 Fourier Components

Experiments with stimuli that contain several discrete frequency components may be considered as the crucial tests of the frequency filter (or Fourier analyzer) hypothesis. According to this hypothesis, the components should be explicitly encoded in the neuronal responses. Ideally, one should be able to determine each component, as far as it is essential for the perceptual quality of the stimulus, from the strengths of the responses of those neurons that are tuned to the corresponding frequency. The experiments with bars discussed above are of course tests of this kind; however, since single bars have a continuous, very broad spectrum, nearly every spatial frequency filter should respond to them, and the discrimination between the Fourier filter hypothesis and alternative (e.g., line detector) concepts is therefore not so easy, at least on a qualitative level (cf. Albrecht et al. 1980; and the discussion in Stork et al. 1982). Bar and grating responses have to be compared quantitatively via spatial frequency tuning, as in the experiments described above. Experiments with stimuli containing a few frequency components with relatively high energies are intuitively more conclusive.

Maffei et al. (1979) have studied responses to gratings of various profiles. They demonstrated that the responses of both simple and complex cells are determined by the fundamental component when the grating periodicity is near the cell's preferred periodicity. The cells responded to a square wave as if it were a sinusoid with the amplitude of the square wave's fundamental and did not respond to a square wave without the fundamental, although this grating still had the same periodicity (that was optimal for the cell). Responses to gratings with somewhat longer periodicities than that preferred by the cell were determined by higher harmonics; e.g., a simple cell showed three response peaks to a cycle of a drifting square wave if the third harmonic fell within the cell's frequency pass band (Fig. 1B).

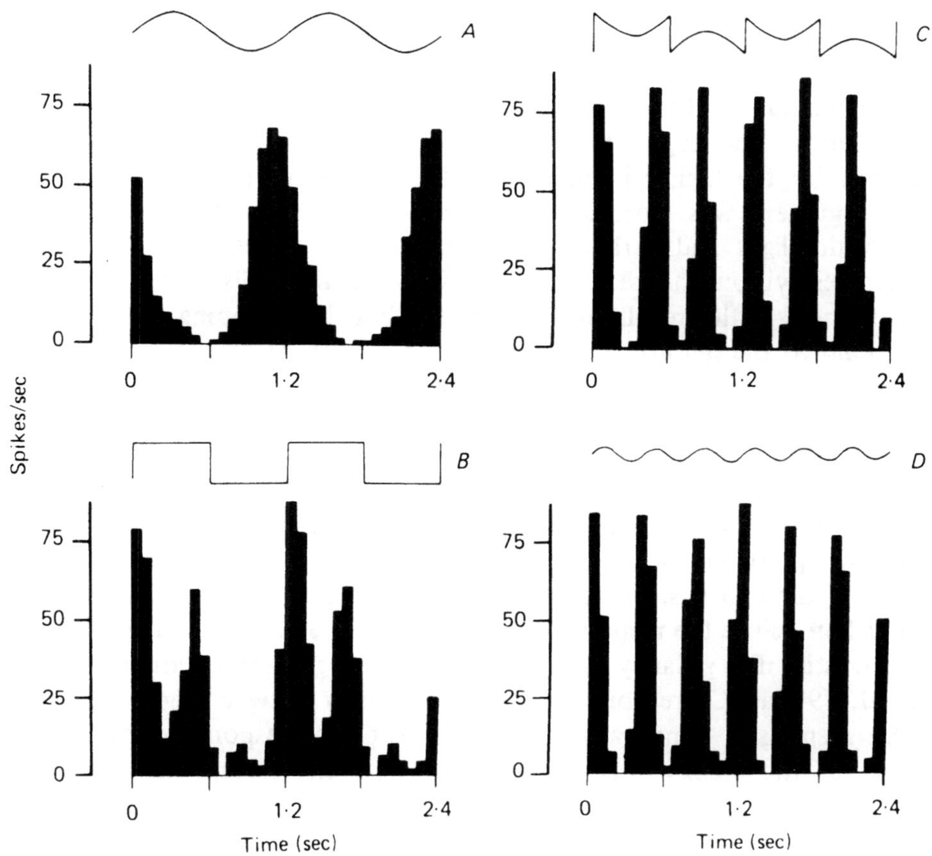

Fig. 1 A–D. The responses of a simple cell in cat striate cortex to drifting gratings of various luminance profiles, as indicated above each histogram. The histograms represent the averaged time course of the cell's firing rate. The spatial frequency of the square wave grating (**B**) was chosen so that the cell responded to its fundamental (**A**) as well as its third harmonic component (**D**), while fifth and higher harmonics were outside its frequency range. Grating (**C**) was a square wave from which the fundamental component had been subtracted. The sine wave produced one response peak per grating cycle (**A**), the square wave three (**B**), although both gratings had the same periodicity. The response to the missing-fundamental grating (**C**) equalled the response to the third harmonic alone (**D**), although the two gratings had different periodicities. Under certain assumptions (see text) these results follow from linear filter theory. (Reproduced, with permission, from Maffei et al. 1979)

When the fundamental of that grating was then removed, the response became similar to that produced by a sine wave grating equal to the third harmonic component (Fig. 1C and D). Pollen and Ronner (1982) have confirmed and extended this result by showing that square wave spatial frequency tuning curves of simple as well as complex cells often show a separate secondary peak at the third harmonic frequency. Thus the responses conformed here entirely to the linear filter prediction.

What were responses to square waves of very low spatial frequency like? Here the single edges of square wave gratings acted independently, and when the fundamental of such a grating was removed the responses did not change, in agreement with the linear filter hypothesis, since the fundamental of these gratings was well below the cell's low cut-off frequency. One might think then that the neuron's response was determined by the higher harmonics of the square wave grating. However, the argument was more subtle; the response was said to be "*correlated* with the light and dark edges . . . because the edges contain high frequencies within the range of sensitivity of the cells", but, at higher frequencies, "the results are interpreted best by assuming that . . . cells respond to the harmonics . . ." (Maffei et al. 1979). Does this mean that the edge-related responses can be predicted from the frequency tuning?

Figure 2 shows responses of a simple cell to gratings of 0.12 cycles per degree. (I could not work out what its optimum frequency was, but probably it was at least four times higher.) The response to the drifting square wave grating, and that to the grating without fundamental, show almost equidistant peaks of equal amplitudes, one corresponding to each edge of the gratings. Obviously the above definition of linearity does not hold here, since the responses cannot be interpreted as showing inversion when stimulus polarity is inverted, neglecting rectification (Movshon et al. 1978a). Corresponding to the similarity between light and dark edge responses, temporal Fourier analysis of the responses showed that most of the energy was concentrated near the second harmonic and some at the fourth. Very little or nothing was found at the odd harmonics (three, five, seven, and nine) which constitute the Fourier spectrum of a square wave grating. These observations seem to disagree with the linear filter model. Would it be possible to understand them if time dependence and rectification were included explicitly in the model? Note that the example discussed was a simple cell. Similar indications of nonlinearity, perhaps more pronounced, were observed in complex cells.

Regarding the long-standing controversy between feature detector and spatial frequency analyzer models, Maffei et al. believe that their results show that "both hypotheses can be phenomenologically correct".

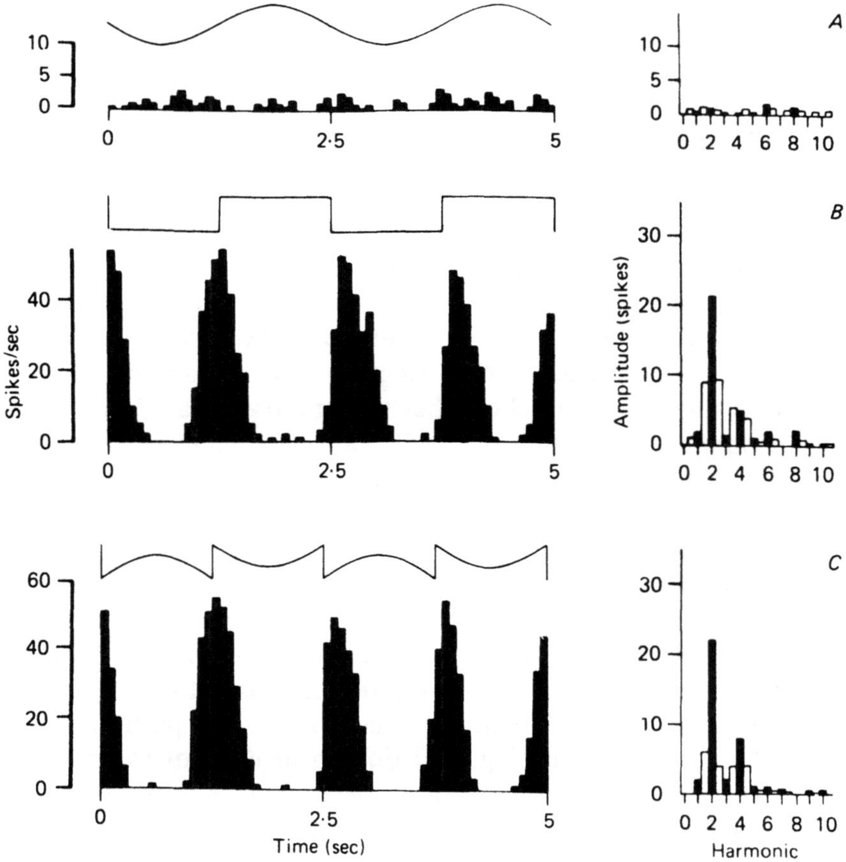

Fig. 2 A–C. The responses of a simple cell to gratings with cycles much larger than the width of the receptive field (8.3° versus 2°). Similarly to Fig. 1, responses to a square wave grating (**B**), its fundamental component (**A**), and the missing-fundamental grating (**C**) are shown. Spectra obtained by temporal Fourier analysis of the responses are shown on the *right*. Responses components at harmonics of the drift frequency are shown in *black*. As expected from the cell's spatial-frequency tuning, the cell did not respond to the sine wave grating since its frequency was too low, and responded to square-wave and missing-fundamental gratings similarly, since the two differed only in a frequency component which had no effect by itself. However, the responses to positive and negative going edges are similar, and thus cannot be interpreted as the positive part of a linear filter response. (Reproduced, with permission, from Maffei et al. 1979)

An interesting point of the study of Maffei et al. (1979) was that the neural responses did not change when the fundamental was removed from a square wave grating of low frequency. This parallels perception, where such gratings may be indistinguishable (Campbell et al. 1978). Maffei et al. also point out a difference from perception, namely that the periodicities of missing fundamental and square wave gratings are correctly perceived, even at low frequency, where the neurons responded with double frequency. When grating frequency was near preferred frequency, however, the cells failed to respond when the fundamental was removed.

The perception of the missing fundamental square wave grating is similar to the phenomenon that spatial "beat" frequencies are readily perceived, although they are not in the Fourier spectrum of the stimulus. Albrecht and De Valois (1981) have studied responses to beat patterns obtained from superposition of three sinusoids with frequency ratios 4:5:6. The resulting pattern is a sinusoidal grating of frequency 5 whose amplitude is modulated with a frequency of 1. Henning et al. (1975) had used this pattern to show that the proposed spatial frequency channels (Campbell and Robson 1968) are not independent. They had found that in a detection task the pattern was masked effectively by a sinusoidal grating of the beat frequency, and conversely that the simple sinusoid was masked by the beat pattern. This was quite surprising because the spectra of the beat pattern and the simple grating are separated by two octaves. Albrecht and De Valois found no response to the beat pattern in cells that preferred frequencies near the missing fundamental. However, simple cells with preferred frequencies in the range of the components (4, 5, and 6) responded to the drifting beat pattern, as expected from their frequency tuning, and Fourier analysis of this response also showed a strong component at the beat frequency, explained by the cells' output nonlinearity (partial rectification). They also found some interaction when the two stimuli, the simple sinusoid and the beat pattern, were presented superimposed. Both the responses to the simple grating (in cells tuned to its frequency) and the responses to the beat pattern (in cells tuned to its frequencies) were diminished by the presence of the other pattern (whose frequencies were outside the tuning). Such negative interaction (average, 25%) was observed in about half the cells, enhancement was observed less often. Finally, it was shown that the interaction effect was not specific to the beat pattern, it could be equally obtained with just one of the high frequencies with corresponding contrast. Similar interaction effects between two gratings of widely different frequencies have been observed in other studies (see De Valois and Tootell 1983). Albrecht and De Valois (1981) believe that their results have not only failed to verify the model proposed by Henning et al., but have also provided a different explanation for those psychophysical results. Firstly, the interactions between beat pattern and simple grating observed in neuron responses, they say, parallel the psychophysical masking effects. Note, however, that Henning et al. did control for specificity; no masking was obtained at the fundamental by just one of the high frequency components, and vice versa, confirming earlier psychophysical studies and at variance with the neuronal data. Secondly, Albrecht and De Valois (1981) explain the perceived periodicity of the beat pattern by the pattern of periodic excitation in striate cortical cells, which, due to rectification nonlinearity, contains a Fourier component at the frequency of the missing fundamental. In essence, this explanation defers the detection of the periodicity to a later stage of processing (as did the model of Henning et al.). Unfortunately, one cannot explain the perception of periodicity of the missing fundamental square wave grating in the same way because, in this case, the activity pattern of cortical cells does not contain the proper Fourier component (Maffei et al. 1979, see Fig. 2 above).

Stimuli without low frequencies may be constructed from various patterns, including geometrical illusion figures (Carlson et al. 1984). It was shown that these patterns are perceptually similar to the originals and may even produce the same illusions, e.g., in the length of the arrows in the Müller-Lyer figure.

Two dimensional patterns provide some interesting tests of the linear filter hypothesis. A checkerboard has a Fourier spectrum to stimulate the imagination. In contrast to a square wave grating, which has all its spectral components aligned on one axis of the frequency plane (the axis perpendicular to the grating stripes), the checkerboard has no energy on the principal axes; its fundamental components lie on the diagonals, corresponding to gratings oriented at 45° to the edges (which can be easily verified by squinting), and higher harmonics have still other orientations (Kelly 1976; De Valois et al. 1979). De Valois et al. found that simple as well as complex cells did indeed respond maximally when the diagonals of the checkerboard, i.e., the fundamental components and not the edges, were parallel to the preferred orientation. Responses related to the higher harmonics were also demonstrated. Those obtained with a three times coarser checkerboard in fact peaked near the orientations of the third harmonics, at 18° to both sides of the edge orientations.

The fact that both simple and complex cells responded to the checkerboard stimulus as expected from its fundamental components which are two superimposed gratings crossed at 90° should be compared with results of other studies showing that responses suffered strong "cross-orientation inhibition" when an orthogonal grating was superimposed on an optimally oriented stimulus (grating or one-dimensional noise; Petrov et al. 1980, Morrone et al. 1982). Morrone et al. found that this inhibition was stronger in simple than in complex cells. By contrast, De Valois et al. (1979) found that simple cells responded to plaid patterns, i.e., two superimposed crossed square wave gratings, just as to a single such grating. Similarly, Movshon et al. (1985), moving plaid patterns in a diagonal direction, obtained orientation tunings as predicted from the two components, i.e., the cells responded to the grating of their preferred orientation and the fact that a second, orthogonal grating was superimposed and moving sideways did not much disturb them, although some inhibitory effect (25%–40%) was observed. (The point of this study is the differing behavior of MT responses and their relation to movement perception, which are not discussed here.) At variance with Petrov et al. 1980 and Morrone et al. 1982, these results argue for independent processing of Fourier components. On the other hand, responses to first harmonics may not be a strong argument for linear filtering, since the nonlinear, periodicity-sensitive cells of von der Heydt et al. (1986a) (see below) also signaled the fundamentals of checkerboards.

In the experiments of De Valois et al. (1979) the preferred grating frequency of a cell was determined first, and a check size equal to the bar width of this grating was then chosen to record orientation tuning with the checkerboard. In terms of the edge detector philosophy, the stimulating edges were therefore very short, while the diagonals were long. At larger check sizes, however, the responses must eventually correlate with the edges, as the receptive field is limited (MacKay 1981). However, here the two hypotheses become indistinguishable, since changing to larger check sizes (relative to the receptive field size) shifts higher harmonics into the cell's frequency passband, whose orientations are closer to the edge orientations.

De Valois et al. based their analysis of simple cells on the time course of the responses to the drifting pattern and used temporal Fourier analysis to isolate the responses to the first and third harmonics. This means, of course, that these components were not explicit in the responses but still had to be detected from the response patterns (cf. the above discussion of experiments with missing fundamental gratings). However, this criticism does not apply to the results from complex cells, which responded with an unmodulated increase in firing and peaked at the orientations of the third harmonics (biased towards that of the fifth) (Fig. 12 of De Valois et al. 1979). Here, the harmonics must have been sensed by neural spatial integration. This result is really surprising. It implies summation over a considerable distance, either along the receptive field orientation with positive definite weighting, or perpendicular to it with periodically oscillating weighting, or a combination of both.

More about neural filtering of two-dimensional patterns is to be expected (Grosof et al. 1985).

3.5 Some Nonlinear Phenomena

Some studies have specifically addressed nonlinear phenomena in cat striate cortex. Interactions between two spatial frequency components in drifting gratings were studied by De Valois and Tootell (1983). One was fixed at the optimum, and the other was varied and presented with various spatial phase relations. Strong interactions were found to be common, if not the rule, particularly in simple cells. Consider too, in this context, the effects of textured background on responses to localized test stimuli (Maffei and Fiorentini 1976; Nelson and Forst 1978; Hammond and MacKay 1981; Morrone et al. 1982; Nelson and Forst 1985). All these studies demonstrated nonlinear interactions between background and test stimulus. In general these effects were stronger in simple than in complex cells. Complex cells show nonlinearities by definition. Spitzer and Hochstein (1985a) studied the harmonic distortions in simple and complex cell responses to stationary, counterphase-modulated gratings as functions of spatial frequency and phase.

Extreme nonlinearity was found in *temporal* summation of simple and complex cells (Tolhurst et al. 1980). Responses to stepwise contrast changes were impulse-like, and in general much shorter than linearly predicted from temporal frequency-response functions obtained with sinusoidal modulation. Failure of superposition in simple cell responses was demonstrated with gratings modulated at two different time frequencies (Dean et al. 1982).

Another type of nonlinear behavior is the phenomenon of contrast adaptation. This is particularly evident when grating stimuli are used. During prolonged stimulation, responses of many cortical neurons decline progressively over 1 min or so, and after such stimulation their contrast thresholds are elevated and recover gradually in the absence of stimulation (Maffei et al. 1973). The behavior parallels the psychophysical effects of contrast adaptation. Recent studies interpret it as a mechanism for maintaining a high differential sensitivity in channels with limited dynamic range ("constructive gain control", Ohzawa et al. 1982; Albrecht et al. 1984). When contrast response characteristics of a cell were determined in several series, each presenting only contrasts within a limited interval instead of the whole range, a gamut of parallel curves was obtained, one for each contrast interval. The perceptual adaptation effects are thus considered to be a consequence of the computational needs for processing images with varying ranges of contrast, rather than epiphenomena of a general property of neuronal responses. Note that the effectiveness of a stimulus in driving the cells, and not just its contrast, is significant for the magnitude of the adaptation effect (Vautin and

Berkeley 1977; von der Heydt et al. 1978; Movshon and Lennie 1979). Besides stimulus contrast, orientation, direction of movement, and spatial frequency have been shown to be important, and the various factors interact. The direction-specific adaptation effect, for example (von der Heydt et al. 1978), cannot be easily interpreted in terms of contrast sensitivity changes. In any case, the gain-setting concept allows the inclusion of some nonlinear effects in a quasi-linear model.

3.6 Texture Discrimination and Periodicity Detection

The investigation of the texture discrimination capabilities of cortical neurons is still at its beginning. Since it has been found that complex, but not simple, cells respond to moving random dot patterns (Hammond and MacKay 1977; but see Skottun et al. 1985), the responses of (some) complex cells might be thought to be related to the perception of texture. Nothdurft and Li (1984, 1985) studied the encoding of texture difference in neurons of cat striate cortex. They used patterns made up of evenly spaced short lines; for example, one of their stimuli consisted of a circular field of "needles", all of the same orientation and length, surrounded by a field of needles of the orthogonal orientation. The needles in the two regions could have equal or different luminances. By movement of the stimuli in a raster fashion, receptive fields were made to scan these patterns, and the responses were displayed as maps (Creutzfeldt and Nothdurft 1978). With coarse textures, orientation differences, but not luminance differences, caused differential firing rates; with fine patterns, the reverse was seen in simple and in complex cells. In other words, when the lines were spaced widely their orientations were resolved but luminance made no difference (because of response saturation), and when the patterns were dense enough for differences in mean luminance to be detected, orientations became unimportant. Luminance differences caused elevated firing at the border between the textures, while orientation differences caused increased or decreased activity all over the textured regions. Thus, pure texture borders were not made explicit. Complex cells detected luminance differences in much coarser patterns than simple cells. In this case, complex cells often signaled the circular border all around, i.e., without regard to its orientation, a peculiar observations. Nothdurft and Li (1985) conclude that cells in area 17 "cannot distinguish whether the contour is part of a border between larger areas or whether it is part of the texture within such an area itself".

Grating Cells. In the early days of the cortical Fourier analyzer hypothesis interest focused on the neurons' spatial frequency selectivity, and their preference for grating stimuli. Curiously, in monkey striate cortex, cells have now been found that respond selectively to gratings of certain spatial frequency but are not spatial frequency filters (von der Heydt et al. 1986a). Such cells can detect periodicity in gratings of various profiles and in checkerboard patterns but do not signal Fourier components, such as higher harmonics of square wave gratings (Fig. 3) or spectral components of bars and edges.

They did not respond to aperiodic stimuli, even of high contrast (although these contained considerable spectral energy in the cells' pass-bands) and were also unresponsive to square wave gratings of one-third their preferred frequency, even though the third harmonic components of these gratings, presented alone, elicited strong responses. The narrowest tuning to spatial frequency (0.43 octave bandwith at half-height), and the highest peak frequency (20 cycles per degree), were found among these cells. All of them responded best to gratings of many cycles (4–14). Thus, frequency selectivity and grating preference do not imply linear filtering. Such "grating cells" made up about 5% of the units recorded in V1 and 1.5% in V2, in samples of about 500 from each area. It was concluded that these cells signal periodicity and orientation of certain textures (e.g., the fur of an animal). The reason why these specialized detectors are found in the primary visual area must be that the high demands on spatial resolution and the redundancy in a point-by-point representation call for a more efficient encoding of texture at an early stage. There were interesting parallels to perception. Most grating cells responded to gratings without regard to phase and without direction preference, in parallel to the observations that directions of phase shifts cannot be discriminated (Badcock and Derrington 1985) and movement may not be perceived (Campbell and Maffei 1979) in high-frequency gratings. The fact that grating cells are not excited by higher harmonics may explain why, in some experiments, square wave gratings did not produce contrast adaptation at third harmonic frequencies (e.g., Nachmias et al. 1973).

Von der Heydt et al. (1986b) found that the responses of grating cells could not be reconciled with a linear filter model, nor with a summation-to-threshold model. The problem with these models is that they predict responses to aperiodic stimuli and to the harmonic components of nonsinusoidal gratings. If one eliminates these responses by assuming an output threshold then insensitivity to low-contrast gratings results, but the grating cells did respond at low contrast. On the other hand, it was shown that a nonlinear periodicity detector model based on

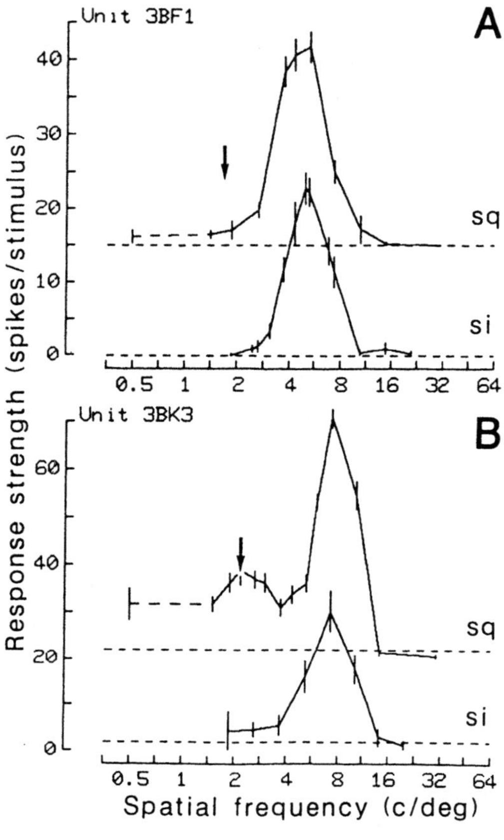

Fig. 3A, B. The spatial frequency dependence of the responses to square wave (*sq*) and sine wave (*si*) gratings of two cells in monkey striate cortex. The square wave tuning curves have been offset vertically by 15 and 20 units respectively. ***Dashed lines*** indicate levels of spontaneous activity. The unit in **A** was a "grating cell", i.e., a cell that responded selectively to gratings but not to aperiodic patterns. The unit in **B** was a complex cell that responded to gratings as well as to bars and edges, with a preference for dark bars. The two units showed similar spatial frequency tuning with sinusoids, but with square waves the grating cell did not respond at one-third of the optimum frequency, where it should sense the third harmonic according to spatial frequency filter predictions. The "normal" complex cell showed the predicted extra peak. ***Arrows*** mark frequencies and amplitudes of the extra peaks predicted from the cells' sine wave tuning curves and the Fourier spectrum of the square wave stimuli. Further tests showed that the failure of the grating cell to signal the third harmonic component was not a matter of high contrast threshold, since it responded strongly when this component was presented alone (12 spikes per stimulus). (From von der Heydt et al. 1986a)

simultaneous detection of three slopes of alternating signs in the luminance distribution can account for the behavior of a typical grating cell, and a similar model with four or five slope detections for the cases of narrow tuning.

3.7 Length Summation

Linear spatial filtering, of course, implies linear "length summation". Indeed, the notion of length summation in cortical receptive fields stimulated the Fourier analyzer hypothesis (Pollen et al. 1971). To some extent length summation was tested in the experiments with checkerboard patterns described above. The studies on spatial frequency tuning explicitly or implicitly made assumptions about length summation. In the Gabor function model, weighting along the length dimension is usually modeled by another, multiplicative Gaussian function (i.e., no oscillations in this direction are assumed) (Kulikowski et al. 1982; for discussion of alternatives see Daugman 1980, 1983; and Webster and De Valois 1985). Hubel and Wiesel (1962) suggested that a row of concentric (LGN) fields provide the input for a simple cell. Webster and De Valois have compared these models with the orientation and frequency tuning in cat striate cortex.

On the other hand, one indication of nonlinearity has often been described (e.g., Bishop and Henry 1972; Schiller et al. 1976a): cells that respond well to long bars sometimes show "negative" response fields, i.e., in order to evoke a response, a bar has to extend in length beyond the receptive field center on both sides. No response is obtained, for example, when an optimally oriented bar covers half of the receptive field on one or the other side. The bar can be of high contrast. Simple and complex cells may respond this way.

Since this observation demonstrates particularly well the preference for elongated stimuli, it has stimulated ideas about feature detection. Marr and Hildreth (1980), for example, have postulated nonlinear operations similar to a multiple logical AND in the longitudinal direction for detection of oriented zero-crossing segments (but see a critical analysis of this assumption in Torre and Poggio 1986).

The linearity of length summation, and in particular the suggestion of Marr and Hildreth, were recently tested in a study by Schumer and Movshon (1984), who used gratings of various combinations of length and contrast. With these stimuli, the suggested nonlinearity should be apparent, as the cells should respond only to some minimum lengths, whereas the assumption of linear summation would predict responses to short stimuli too, if they have high contrast. The experiment showed that the cells' responses were determined by the product of contrast and length over large ranges of either variable, in accordance with the linear summation model, but at variance with the AND model. Observations of nonlinear length summation in receptive fields such as the "negative response fields" were explained by a high output threshold (summation-to-threshold model).

I think this result should not be taken as the ultimate conclusion about length summation in simple cells. First, it is not clear whether one of the cells tested actually showed a negative response field. Secondly, gratings may be different from edges in such length summation tests. Heggelund et al. (1984) found different length summation in the ON and OFF regions of the same cell, for the excitatory as well as the inhibitory effects. All these effects are pooled in some way when gratings are used, as in the experiment of Schumer and Movshon. The suggestion of Marr and Hildreth, after all, referred to edge detection.

Certainly, one should not generalize the findings of Schumer and Movshon (1984) to mean that length summation is linear under all conditions. Morrone et al. (1982) compared a two-dimensional visual noise pattern with the one-dimensional pattern obtained from it by integrating (smearing) along one axis. The two-dimensional pattern evoked much weaker responses than the one-dimensional pattern, particularly in simple cells. Thus, neuronal and physical length summation were quite different here. Hammond and MacKay (1983), inspired by the experiments of De Valois et al. (1979) with checkerboard patterns, set out to investigate the question whether positive and negative edge contrasts are summed algebraically along the length in receptive fields of simple cells (and complex cells, Hammond and MacKay 1985). Cancellation of contrast would explain why the cells do not respond to the edges of checkerboard patterns (MacKay 1981). They found, quite unexpectedly, that contrast reversal within the receptive field often had a very strong, nonlinear effect. The responses to a light bar, for example, dropped to zero when the bar was extended with dark bar segments. Summation was nonlinear, because the drop, as a function of segment length, was steeper than the slope of the light-bar length summation curve, i.e., the contrast reversal did more than just reverse the effect of length summation. Similar suppression effects were also observed when a segment in the center of a bar was inverted. Again, strong effects (70%) were produced by relatively short (e.g., 0.5°) segments. Longitudinal weighting functions for suppression, found by placing an inverted segment at various positions along the bar, could be broader than the range of length summation. Related experiments are those of Heggelund et al. (1984), in which a short and a long bar were superimposed and switched on and off in counterphase. The short bar reduced the excitatory responses to the long bar. Heggelund et al. (1984) used two light bars on a homogeneous background, whereas Hammond and MacKay (1983) used a technique of displaying light and dark bars in a random pattern of light and dark dots.

Also the psychophysical result that, in random dot patterns, elongated regions of higher density are not detected better than square regions (Barlow 1978) contradicts the linear length summation concept.

In the search for a functional interpretation, one may compare these findings with results in monkey V1, where cells could not be activated by bars or edges with gaps (so-called illusory contour figures) even when the gaps were narrower than the range of length summation (von der Heydt et al. 1984; Baumgartner et al. 1984; the further results of these studies will be discussed below). Furthermore, Peterhans et al. (1984, 1986) found that simple and complex cells in V1 nearly always responded less to dotted lines than to continuous lines, and less to edges of gratings than to edges of homogeneous fields of equal mean luminance. The effect was often dramatic, responses of some cells dropped to zero when a continuous light bar was replaced by a stippled bar of equal mean luminance with a stippling frequency as high as 8 cycles per degree Even more surprising, a single gap as narrow as 4 min arc, in a bar, could abolish the response. By contrast, the longitudinal weighting function was broader than 30 min arc.

These results and the findings of Morrone et al. (1982) and Hammond and MacKay (1983) suggest the following interpretation in terms of visual function. The suppressive effect of gaps in bars or edges is reminiscent of the longitudinal AND operation in the edge detection scheme of Marr and Hildreth (1980), although the underlying mechanism may be different. It could serve the function that only zero-crossing segments of certain length are encoded, while parts with discontinuities are rejected (cf. conclusions of Morrone et al. 1982). The reason for this strategy may not be obvious, since it is often assumed that perception of contour requires just the opposite, namely the continuation of lines or edges across discontinuities. This seems to be demonstrated by the famous illusory contour figures (see example in Fig. 7). Ullman (1976) presented a model for such "filling in" of gaps. However, apparently, this is not what the striate cortex is doing; simple cells do not respond to interrupted lines or edges as if they were continuous, and often their signals drop towards the end of a line or edge (negative response fields). Thus, the results of the above experiments are compatible with the idea that simple cell signals encode zero-crossing (or possibly some other local property derived from a filtered image) where it forms continuous lines of some minimum length, and do not make terminations explicit, while the points of discontinuity and terminations are encoded in signals of strongly end-stopped cells (hypercomplex cells, Hubel and Wiesel 1965, 1968). This sketch will be filled with more detail in the discussion of contour mechanisms below.

3.8 Conclusions on Spatial Filtering

In summary, the above discussion shows that a large body of experimental results on linear filter operations and nonlinear effects in receptive fields are available. Although the question of linearity appeared to be related to the Fourier analyzer theory in most of the studies, it is of more general interest. I have deliberately pointed out indications of nonlinear processing, although, perhaps, the authors of the studies might not always agree that those incidents should be stressed so much. In fact, most of them have emphasized the compatibility of their results with the linear filter concept and considered the deviations from linearity in simple cell behavior not very important. However, one should not forget here that many studies were obviously motivated, alas, by the intent to defeat the naive feature detector concept. (What could be achieved with bar detectors like a simple cell that signals a light bar in one position and a dark bar in two other positions?). Often one has the impression that the authors felt that a functional interpretation of the observed nonlinearity was not necessary. The truncation of responses at zero-firing level (rectification nonlinearity), it can be argued, is not really important because the signal may be carried in complementary pairs, so that the information required to substract out the effects of truncation would be mathematically available (Pollen and Ronner 1982, p. 114). The "negative maintained discharge level" of simple cells (e.g., Henry et al. 1969; Bishop et al. 1971; Movshon et al. 1978a; Dean et al. 1982; Schumer and Movshon 1984; De Valois et al. 1985), which would spoil this argument, might be assumed to be artificial, perhaps a consequence of anesthesia (though simple cells, and most cells in V1, have low spontaneous activity in alert-monkey experiments too, with no drugs used).

Other studies have tried to integrate the nonlinear effects into the spatial frequency filter concept. Morrone et al. (1982) assumed inhibitory influences of complex on simple cells as the mechanism of "cross-orientation inhibition". Intracortical inhibition was also suggested as an explanation for the interactions between spatial frequency channels; Webster and De Valois (1985) and De Valois et al. (1985) have argued that the nonlinearities may be described by a suppressive surround in the two-dimensional spatial frequency spectrum.

I would suggest that incidents of nonlinear processing be investigated more by themselves, and not just as deviations from a hypothetical linear model. Describing nonlinear behavior of cortical neurons as deviation from linearity is complicated and may not lead to an understanding of their function. On the one hand, one gives up the generality and simplicity of the linear formalism, on the other, functional interpretations of the added nonlinearities are sought ad hoc, e.g., sharpening of spatial frequency tuning. Although Ginsburg (1982) reminds us that models are always approximations and that it is not wise "to dismiss a particular model because it is incomplete (for example, failing one particular test)", we should not forget also that a system that proves linear in some tests and nonlinear in others is, after all, a nonlinear system. The strategy of first approximating the data with a linear model and then trying to improve the approximation by adding nonlinear elements may be adequate for some psychophysical experiments where one may not be able to separate several stages of processing. In neurophysiology, where every step is, in principle, accessible, one should rather try a direct approach. Rather than regarding simple cells as "essentially linear", one could perhaps find out which of them are nonlinear and why.

Linear transformations can be preliminary stages in visual processing. Some of them are theoretically reversible, e.g., the center-surround transform used for calculation of lightness (Horn 1974), or the predictive coding used for efficient transmission of pictures from satellites (Gonzales and Wintz 1977). This makes their interpretation ambiguous (see also discussion of color-opponency below). Many nonlinear models require an initial stage of linear filtering (e.g., Marr and Poggio 1979; Marr and Hildreth 1980).

On the other hand, one can expect nonlinear features of cortical receptive fields to point to mechanisms of visual encoding that are more principle in nature and may thus hope to find here the keys to cortical function. For example, the finding that grating cells, as a consequence of nonlinearity, are not activated by certain Fourier components of stimuli narrows down the range of possible functions considerably, indicating a role in texture discrimination. Similarly, the nonlinear effects in simple cells of gaps (or contrast reversals) in bars and edges are suggestive of a function such as checking continuity of zero-crossing segments. The need for nonlinear networks for specific visual tasks has long been recognized in insect studies (Reichardt 1961, 1986). While theoreticians demonstrate new possibilities for nonlinear computations at the synaptic level (Poggio and Torre 1978; Koch and Poggio 1983), physiologists of mammalian visual systems should no longer try to get along with linear descriptions and with the too simple summation-to-threshold model.

4 Color

4.1 Perception

I have found it particularly difficult to review the advances in the physiology of color vision, since several laboratories have recently published their results more or less simultaneously, and therefore without referencing each other, and each apparently with a different idea of what the physiology of color vision is. Therefore, before we proceed, it may be a good idea to say a few words on the word color.

This word has of course several meanings, but in the field of color vision it is useful to reserve it for the perception. The word color and the color names are often used as if denoting properties of objects, as in common language, but they are also used sometimes, particularly in scientific language, to denote the spectral composition of the light that reaches the eye from the surface under consideration. These two types of usage may be called the naive and the scientific errors. Unfortunately, the more confusing one, the scientific error, is most widespread in the physiological literature. It suggests that color perception is determined by the spectral composition of the stimulus (or its trichromatic values) which is in general not true. From time to time, this appears as recent insight (Zeki 1983a, 1984), but these relationships were explained some time ago with perfect clarity (Hering 1878)[2]. (Note that this insight is not recent either.) In general, the perceived color of any part of a scene depends also on the light coming from the rest of the scene. This is well-known, for example, as the phenomenon of color constancy, showing that illumination, and thus the composition of the light that each object sends to the eye, can change widely without

---

[2] Zeki quoted a didactic statement in which Hering cautioned the reader that, in color experiments (those of his time), familarity effects can influence color judgments if color is not being particularly attended to. This remark is purported by Zeki to show that Hering recurred to memory and experience as explanations of color constancy and that Hering, like Helmholtz, was unable to take the "gigantic step needed to understand that colour vision is a creation of the brain" (Zeki 1984). If a gigantic step is needed, then Hering was a giant indeed.

affecting the colors of the objects. The opposite, that the color of an object can change drastically without any change in the light reflected off it, can also be demonstrated easily; if, for example, a yellow disk is shown by projecting filtered light onto a screen in the dark, and the area surrounding the disk is then illuminated with bright, unfiltered light, the disk turns brown. What has become more generally accepted, perhaps, since the demonstrations of Land (1959a, b), is that the independence of color from local wavelength composition is not a small effect and is essential for normal vision. Color is not the result of some small correction on the local trichromatic values, but requires elaborate calculations taking account of the whole image. Curiously, the "naive error" is not so far from the truth, since the visual system achieves a high degree of color constancy, i.e., color approximates to a measure of object property. Incidentally, there is evidence that "color contingent" aftereffects are really wavelength contingent (Thompson and Latchford 1986). The "pure" or "unique colors", on the other hand, seem to be really colors and not wavelengths, since they are subject to induction effects (Akita et al. 1964; Oyama and Hsia 1966).

Note that black and white are also colors in this sense, and consequently, Hering's theory deals with three pairs of opponent colors (colors that cannot be perceived simultaneously since they constitute the two directions of variation of one and the same signal).

The use of the word "color" with different meanings is not necessarily confusing if it is used consistently within a paper (and if the readers are flexible enough to make such changes between papers). For example, the dimensions orthogonal to the luminance axis in trichromatic space may be called the color axes; in this sense, one often speaks of isoluminant colors. However, in a discussion of the perception of white and its neural representation, saying that a hypothetical white encoder "would respond to various colors plus their complements" (Vautin and Dow 1985) is confusing because of the conflicting use of the terms "color" and "white".

Retinex Theory. Land and McCann (1971) have proposed a theory that explains many of the color phenomena, and allows fairly accurate prediction of color perception in complex displays. It is possible, as critics claimed, that this theory did not achieve a better quantitative description of color vision than formulas that had been derived from experiments before; however, as Marr (1982) pointed out, it is important to note that Land and McCann's approach is not merely a description of color perception; its essential part is a formulation of assumptions about the physical world on which vision might be based. It is thus a theory of color vision. The problem is how reflectance of object surface could be determined under conditions of unknown, variable, and inhomogeneous illumination, and without a reflectance standard. The proposed solution is based essentially on the assumption that those intensity changes across the image that are due to illumination are gradual, whereas those due to differences in reflectance are abrupt. The theory shows how the effects of illumination can then be discounted. Such a theory is particularly valuable for the physiologist. First, it specifies precisely what has to be computed, and thus leads to specific suggestions of physiological mechanisms (see, for example, McCann et al. 1976). Secondly, it states explicitly the conditions (the "assumptions about the physical world") under which this computation gives the desired results. Thus, the predictions of the theory, and their deviations from experimental results, can be traced back to those conditions, which may lead to improvements in the theory, or the formulation of new and better theories (Marr 1982). For recent summaries of Land's Retinex theory, computational methods, and related physiological results see also Land (1983, 1986), Zeki (1983a), and Livingstone and Hubel (1984a). For an expalanation of its principles and psychophysical evidence, I found McCann et al. (1976) particularly valuable.

As to the question where the long-range interactions postulated by the Retinex theory occur, observations in a split-callosum subject point to a cortical site (Land et al. 1983). This observer judged the color of a test field that he fixated, while a laterally adjacent larger area was being illuminated with one of two different light compositions. In normal observers, these illuminations produced different colors of the test field. For the split-callosum observer, this was the case when the variously illuminated area was on the side stimulating the hemisphere that reported the color of the test field, but not when it was on the other side. Thus, the color of the part of the test field seen by the verbal hemisphere was apparently not influenced by the illumination seen by the other hemisphere. On the other hand, Pöppel (1986), in a hemianopic patient, has demonstrated a color induction effect that appears to be a retinal phenomenon. In this experiment, color induction occurred in the afterimage. The hemianopic observer reported an induction effect which depended on a part of the stimulus that had been presented within the scotoma. Thus, stimuli that looked identical to the observer induced different colors in the afterimage. The interaction occurred here over a long range, as in the experiment of Land et al. (1983), but color differences were induced from a part of the retina having no cortical representation.

4.2 Neuron Studies

The first physiological studies that used the term "color" in its psychological meaning, and consequently distinguished between wavelength selectivity and color coding (and to my knowledge still the only ones), are those of Zeki (1983a, b). Once the power of this basic distinction is recognized, it becomes difficult to understand those earlier discussions about color specificity in cortical areas, in which wavelength preference or selectivity, bandwidth of tuning, or spectral opponency, in a variety of definitions, were used as criteria for determining how many cells are "color selective" in different parts of the visual system.

The uncertainty about what is to be regarded as a step in processing towards the perception of color has influenced the discussion over the question whether area V4 is somehow specialized for color. According to Zeki, there is an exceptionally high concentration of "color selective cells" in V4 (Zeki 1977, 1978a, b, 1983c), and cells there are more selective than retinal ganglion cells or cells in V1 or V2 (Zeki 1980), a view that has been disputed by others (Krüger and Gouras 1980; de Monasterio and Schein 1982; Schein et al. 1982).

Discrepancies between studies in percentages of color selective cells were attributed to differences in definition (Schein et al. 1982; Andersen et al. 1983; for results in V1 see also Thorell et al. 1984), or to sampling differences, since wavelength selectivity was more frequent in parafoveal than in peripheral V1, while being equally frequent for all eccentricities, but differently concentrated in various parts of V4 (Zeki 1983c). There is common agreement, as Zeki (1983c) points out, that wavelength-selective cells are virtually absent in areas V3, V3A, and MT (or V5 as he calls it), while present (beyond the geniculo-striate pathway) in V2 and V4, and this is most interesting and most important.

Desimone et al. (1985) emphasize the generality of the processing in V4, which they see as a stage in a serial hierarchy of visual areas connecting V1 with the inferotemporal cortex thought to be crucial for visual object recognition. They review evidence, mainly from their laboratories, that neurons in V4 can be sensitive, besides wavelength, to many kinds of information that would be relevant to object recognition. Examples of selectivity for orientation, width, and length of bar stimuli, spatial frequency of gratings, and direction of motion, are demonstrated.

As to the spectral properties of V4-receptive fields, Zeki (1980) found that many cells in V4 were selective for narrow parts of the visible spectrum, and concluded that "the responses of cells in V4 can be said to approximate more closely what may be termed responses to individual colours, or hues, than anything seen in more antecedent parts of the visual system". De Monasterio and Schein (1982), however, who measured spectral response bandwidths and peak sensitivity of color-opponent retinal ganglion cells, found no significant differences between the averaged spectral bandwidths of these cells and those of color selective neurons at higher levels (dorsal lateral geniculate body, striate cortex, and V4) published in other studies. Desimone et al. (1985) discuss these together with more recent results. The vast majority of V4 cells, it is said, showed evidence of spectral sensitivity in that they showed a response that peaked at some wavelength and fell to spontaneous activity at others. However, since the spectral bandwidths of the majority of cells (although narrower than that of cones) was comparable to those of the color-opponent cells in retina and lateral geniculate body (de Monasterio and Schein 1982), and since the cells often gave responses to white light and did not show opponent color responses it was concluded that they are not narrowly tuned "color detectors" (Schein et al. 1982).

Apparently, the narrowness of spectral tuning was taken here as a measure of color specificity, and, as in many studies, stimulus specificity served as an indicator of the function of a cortical area. One wonders, though, what role such "narrowly tuned color detectors" should play in perception.

4.2.1 Detection

Thorell et al. (1984) defined as "color cells" in V1 those cells which respond to isoluminant stimuli. They used bars or edges that differed from the background in color (wavelength composition) but not in luminance, and similarly designed chromatic gratings. Luminance was defined by the behaviorally determined photopic luminosity curve. By comparing, in each cell, the strengths of responses to bar stimuli of a criterion contrast in either pure luminance, or pure color (optimum wavelength; contrast defined by distance in trichomatic space relative to maximum possible distance), they obtained the distribution of a "color response index" of V1 cells. They concluded that at least 79% were color cells, while the others preferred achromatic stimuli; a few cells actually responded exclusively to chromatic stimuli. This result (based on a quantitative criterion) is in contrast to the low percentage of "color-coded" cells reported by Hubel and Wiesel (1968). The discrepancy is explained partly by assuming that evidence of color opponency was a criterion in the earlier study. Thorell et al. further showed that even when various amounts of luminance contrast were added to the color contrast, most cells showed essentially invariant spectral characteristics. This refers to the definition of color cells of Gouras and Krüger (1979), postulating luminance-independent responses. As Thorell et al. point out, the rationale behind this definition is unclear.

Other interesting aspects of the study by Thorell et al. are the responses to color-modulated gratings, showing the same dichotomy of phase-sensitive and -insensitive cells as found with luminance gratings (see discussion of simple/complex classification above), and the spatial frequency tuning functions obtained with the two kinds of gratings. While many cells were similarly tuned, some cells preferred lower frequencies, and were more broadly tuned or low-pass tuned, with pure color variation (differences that might be inherited from the properties of the center-surround color-opponent dLGN cells). As a consequence, the averaged tuning curves over all cells showed less low-frequency attentuation and an earlier decline at high frequencies with color than with luminance, resembling the corresponding psychophysical grating detection curves. The psychophysical difference between contrast thresholds for detection and color recognition in color-modulated gratings of high frequencies was related to a similar sensitivity difference between complex cells and simple cells. Some complex cells without spectral preference could nevertheless detect pure color stimuli.

It is clear that color is defined here by the stimulus coordinates in the elementary color metric, and not as a perception. The study relates neuronal sensitivity to *detection* of color (in this sense). This detection is possible without color perception, just as disparity stimuli in Julesz stereograms can be detected without perception of depth.

4.2.2 Tuning

Vautin and Dow (1985) studied the spectral tuning of cells in the foveal representation. They used trained rhesus monkeys and recorded the responses of cells to stimuli presented in the receptive field while the animal was looking at a fixation spot. The strengths of the ON responses to white light, various monochromatic lights, and to broadband lights (for the purple colors), matched for luminosity, were measured. Color-tuned cells, as selected by a tuning index, were then grouped according to their tuning curves, using a cross-correlation method that clustered together cells with similar curves. This method yielded six groups with peaks of their mean curves at four different wavelengths, corresponding to blue, green, yellow, and red. Excluding two more broadly tuned groups, the authors consider cells of four of those groups as "primary color cells", suggesting that responses of these cells mainly contribute to the perception of color, as described by the opponent color theory of Hering (1964). Real opponency, i.e., mutual exclusion of the paired responses throughout the spectrum, was found only between cells of the red and green groups, while responses of the blue and yellow groups were just linearly negatively correlated. The authors speculate about another stage of opponent combination of yellow-blue at higher levels. It should be noted that opponency is used here for groups of tuning curves, in a different sense than usual.

Vautin and Dow based their analysis on responses from the most sensitive region of each receptive field. They only specify that their stimuli were optimized in shape, and defer the discussion of spatial organization to a later paper. Nevertheless, their tuning curves may be compared with the distribution of color preferences (again for the dominant subregion) of Thorell et al. (1984). Surprisingly, there is not much similarity. The distribution of Thorell et al. (of the midpoints between the wavelengths of half-maximal response) shows a large peak in the orange-red (605 nm) and a smaller peak in the blue (470 nm), and has a deep minimum near 500 nm, whereas one-quarter of the curves of Vautin and Dow peak near 500 nm.

One could ask whether the fact that the tuning curves cluster, with peaks at wavelengths of the primary colors, really means that the signals at this level correspond to the sensations of color, as Vautin and Dow (1985) suggest in their Fig. 12. The shapes of the curves reflect the three cone inputs; in fact, Vautin and Dow show in their Fig. 10 that these tunings might result from linear combinations of the cone signals. Would the tuning curves of primary color cells show induction effects like the "pure colors" (Akita et al. 1964)?

4.2.3 Opponency

Livingstone and Hubel (1984a) have investigated a color system in primate visual cortex that they identified by the cytochrome oxidase dark staining patches (blobs). The study was mainly concerned with the anatomical organization of physiological properties such as sensitivity to orientation and wavelength, both in V1 and V2, and the interconnections between V1 and V2. I shall restrict my consideration to the physiological results and their interpretation in terms of color psychophysics. The paper is also a valuable review of the previously described types of circularly symmetric receptive fields in dLGN and striate cortex. The "blob color system" was characterized by receptive fields with concentric center-surround organization (lacking orientation tuning). Three types predominate, two kinds of double-opponent cells, with either red-green or yellow-blue antagonism, and broad-band cells. (Double-opponent and broad-band cells had been previously described in the cortex.) Double-opponent cells show antagonism between two portions of the spectrum in their center, and the opposite antagonism in their surround. Thus, roughly speaking, they respond to color contrast stimuli, but do not respond to white light of any form, and not to diffuse light of any color (the formulation was more subtle, like "tend to respond poorly or not at all"). The wavelength at which an ON response turned into an OFF response was used to subdivide the double-opponent group into red-green and blue-yellow types, but cone inputs were not explicitly determined. Broad-band cells sum signals from more than one cone type and show no spectral opponency, but center-surround antagonism, with identical spectral sensitivity in center and surround. Compared to the dLGN, the double-opponent type is new, and the center-surround opponent type, most frequent in the dLGN and least appreciated by theoreticians, has disappeared. Two other types of the dLGN (center-only color opponent and "Type IV") were represented in small proportions, also in blobs.

According to Livingstone and Hubel, the spectral properties of cells differed greatly between blob and nonblob regions. In the foveal sample, 70% of the unoriented (blob) cells, but only 39% of the oriented cells were classified as color coded (and even fewer in the parafoveal sample). They emphasize, though, that some orientation-selective cells between blobs were also highly selective in their spectral properties. These results seem to disagree widely with the high proportion of color cells found by Thorell et al. (1984). However, the methods of stimulations are very different in the two studies, and so are the criteria for what is regarded as color coded; sensitivity to isoluminant stimuli on the one hand, and spectral opponency on the other. Also, as the data are presented, it appears that Livingstone and Hubel have used different criteria for oriented and nonoriented cells. Perhaps this follows from the application of their classification scheme for concentric cells. For oriented cells, they say that investigation usually began with white slits, and if the cell was unresponsive, a red slit was tried, then a dark bar, and the other colors. It is not clear how literally this should be interpreted, but I think this procedure is the most plausible explanation for their finding 112 cells preferring white as opposed to only 12 cells preferring black, among 235 tested for color. In my experience, most oriented cells in V1 respond well to light and dark bars, and among those that are selective, light and dark are preferred about equally frequently. In any case, Livingstone and Hubel interpret responsiveness to white bars as lack of color opponency, and thus classified those 112 cells as non-color-coded, and the rest, including the 12 black-bar cells, as color-coded. By contrast, Thorell et al. found double-opponent organization of subregions in 81% of their simple cells (including a few concentrics), while many of their complex cells (that usually lacked color opponency) also showed clear spectral preferences. Thus, there might be more color in the responses of oriented cells than Livingstone and Hubel's study would suggest, and this could be important for theoretical considerations (see below).

The most important aspect of their study is certainly the anatomical one, the demonstration of the concentration of the hypothetical color system in the blobs, its segregation by the intrinsic connectivity of V1 (see also Livingstone and Hubel 1984b), and the discrete reciprocal projections between blobs and the thin stripes in V2. This discovery was accentuated even more by the demonstration of another discrete projection from there to V4 (Shipp and Zeki 1985). Anatomical segregation and concentric receptive field structure, besides the spectral properties, may have convinced Livingstone and Hubel that this is a color system, or better, *the* color system. Cells concerned with chromatic information outside this system, they seem to assume, are involved in other tasks, like form perception, that also need this information, but do not serve the perception of color itself (cf. the "universal color cells" of Thorell et al. 1984).

Livingstone and Hubel (1984a) discuss the Retinex theory, and suggest how their results could be related to it, but the parallels are limited. One suggestion is that double-opponent concentric fields might perform the tasks of computing lightness ratios and thresholding, so that only the abrupt changes would enter further stages of computation. (The responses of double-opponent cells are still wavelength dependent, and thus do not correlate with the perception of color, Zeki 1983a, b.) Another is that the broad-band concentric cells found in blobs may take care of the black-white dimension of color, while blue-yellow and red-green double-opponent cells serve the other two dimensions. [Whereas Michael (178a, b, c, 1979) found double-opponent cells only of the red-green type in the cortex, Livingstone and Hubel (1984a) and Thorell et al. (1984) also found the blue-yellow type, which was, however, less frequent.] This implies a rotation of axes (or other transformation) in trichromatic space, from the original axes given by the spectral sensitivities of the cones. In a quantitative study of chromatic mechanisms in the dLGN, Derrington et al. (1984) found only two distinct types of parvocellular neurons, one that receives opposed inputs from R and G cones, and another that receives inputs from B cones opposed to variously weighted inputs from R and G cones. Some of them had weak chromatic opponency, but a group of achromatic cells was not found. Thus, the type III (or broad-band) cells of Wiesel and Hubel (1966) appeared as variants of type I (center-surround color-opponent) cells. It would require quantitative analysis to see whether the broad-band cells in cortical blobs do indeed form a distinct achromatic group, as implied by Livingstone and Hubel. Thorell et al. (1984), with methods comparable to those of Derrington et al., found a continuous distribution of their color response index between purely color-sensitive and purely luminance-sensitive, with most cells in between, but there was an indication of a small distinct group of purely luminance-sensitive cells. Vautin and Dow (1985) found a small group with flat wavelength tuning.

As Land (1983) remarks, the concentrically organized double-opponent cells, representing a symmetrical operator, cannot signal direction of intensity changes as required by the theory. Thus, a further stage detecting the direction is necessary. On the other hand, the oriented double-opponent simple cells (Thorell et al. 1984) would immediately suit this task, but these would be found outside the blobs only, according to Livingstone and Hubel. The notion of a physiological representation of color information in form of a luminance signal and two chromaticity signals, rather than the three original cone signals, is not new, and Marr (1982) remarks that the theoretical need for this rotation (or transformation) of the coordinate system is not evident (but see discussion below). Livingstone and Hubel invoke evolutionary arguments, and the transition to achromatic vision in dark adaptation, as explanations. Finally, the long-range interactions postulated by the theory are postponed to later stages.

Double-opponent concentric receptive fields (first discovered in the goldfish retina by Daw 1968) appear in many physiological explanations of color vision phenomena. Indeed they seem to be the counterpart, in the domain of color, of the broad-band center-surround receptive fields. The former may explain simultaneous color contrast, just as the latter can explain brightness contrast, and some form of color constancy also follows (Daw 1984). Studies on chromatic Hermann grids (McCarter 1979; Levine et al. 1980) seemed to support this view. However, color vision often differs from achromatic vision. Spatial resolution is lower, and low-frequency attenuation less (see discussion in Thorell et al. 1984). Color Mach bands, and hue shifts in isoluminant Hermann grids, both predicted from the center-surround organization, may be weak or absent (Pease 1978; Oehler and Spillmann 1981). Considering further discrepancies, Oehler and Spillmann concluded that center-surround organization (of double-opponent and broad-band cells) cannot account as easily for the color effects as for achromatic contrast effects.

Single-opponent cells, as we may call the color-opponent center-surround cells, do not lend themselves easily to such interpretations, despite their spectral opponency. "On a simple-minded, intuitive level, a cell that does not respond to such stimuli as a red spot on a green background would not seem to do much for the problem of color contrast" (Livingstone and Hubel 1984a). Although these cells are the great majority, if not all (Derrington et al. 1984) of the cells in the parvocellular layers of the dLGN, Livingstone and Hubel find it unlikely that they are used to build up double-opponent receptive fields, for the obvious reason that their surround has the wrong polarity (e.g., green-OFF for a red-ON center cell, whereas a double opponent cell with red-ON response in the center excites to green-ON in its surround), although they admit that it is hard to believe that all that color information is lost. I think a connection between these two types is not inconceivable. Just as double-opponency can be built up from center-only color opponent cells (as Livingstone and Hubel suggest), by using (**a**) one or a few such cells for the center, and (**b**) many of them, of opposite polarity, for the surround, and adding their signals, one may proceed with single-opponent cells. One would not have to worry much about the wrong surrounds, because these would cancel to a large extent; since they are larger than the centers, the sum of the surrounds of group **a** and the sum of the surrounds of group **b** will be roughly equal in shape, but opposite in polarity. (With suitable weighting functions, cancellation will be complete.) If we recall the distinction between ON, OFF, inhibitory and excitatory effects (section 2.1), an alternative mechanism, in which two groups with the *same polarity* are used, but their signals are *subtracted,* appears more reasonable. In this case, diffuse colored light would excite group **a** and group **b**, and their signals would cancel on subtraction, as actually observed, whereas in the above mechanism, it would activate either group **a** or group **b**, and just inhibit the other, and the cancellation effect in the sum would be small.

It might appear strange that surround inhibition of one stage should be discarded at a later stage. However, opponent processing at early stages may well have reasons other than the computation of color, as discussed by McCann et al. (1976). The light absorbed by cones of any type may easily differ by a factor of 20 between the black and white in a scene, but only by a factor of two or so between the most vivid colors. Therefore, attenuation of low spatial frequency (or predictive coding, Srinivasan et al. 1982) is necessary for the compression of dynamics in the common variation, but would be detrimental for the difference signal. Low-frequency attenuation in luminance, but not in the chromaticities, is precisely what single-opponent receptive fields achieve (De Valois and Pease 1971; Derrington et al. 1984).

4.2.4 Color

Zeki (1983a, b) has analyzed neuronal responses using stimulus conditions that dissociate color (the perception) and spectral light composition, as in the experiments of Land. He used a plate filled with about 20 rectangular shapes of various colors which was illuminated by three projectors providing variable amounts of light in three different bands of wavelengths. By adjusting the three light components, he could make any of the differently colored fields reflect virtually any composition of light. Thus, a single field could be made to reflect each of various light compositions (while maintaining its color), and various fields could be made to reflect the same light composition (while still maintaining their individual colors). The interesting question now was whether the responses of neurons in monkey visual cortex would vary with the composition of light reflected from the area placed under their receptive fields, or with the color of that area, as perceived by a human observer. The answer was, in brief, that there are responses of both types. In V4, many cells responded under various conditions of illumination, whenever an area of a certain color was presented in their receptive field. Their responses could thus not be predicted from the light coming from the single area. Responses of the cells analyzed in V1, by contrast, depended on the composition of the light shone into their receptive fields, regardless of the color of this area as it appeared to a human observer. In other words, the responses of these units in V1 depended entirely on the light reaching a small receptive field, while those in V4 obviously also took into account the light reaching areas outside the proper responses fields. Examples of these two kinds of responses, the wavelength-selective and the color-coded, are shown in Figs. 4 and 5, respectively.

The two studies are not meant to be exhaustive examinations of the color properties in the two areas, and it would be useless to compare their results in detail with those of the other studies described above. However, a number of interesting points are demonstrated by compelling examples. As can be seen in Figs. 4 and 5 (Zeki 1983a), one cannot predict those color-coding properties from a cell's spectral sensitivity; cells with similar wavelength sensitivity curves can show either wavelength- or color-dependent responses. The responses of the former type required a certain amount of light of the preferred wavelength in the receptive field, regardless of which of the differently colored fields of the display was placed under it. Specifically, the cells showing spectral opponency showed this behavior (double opponency was not explicitly studied). It was shown that by changing the illumination these cells could be made to respond either ON or OFF to any of the fields of the

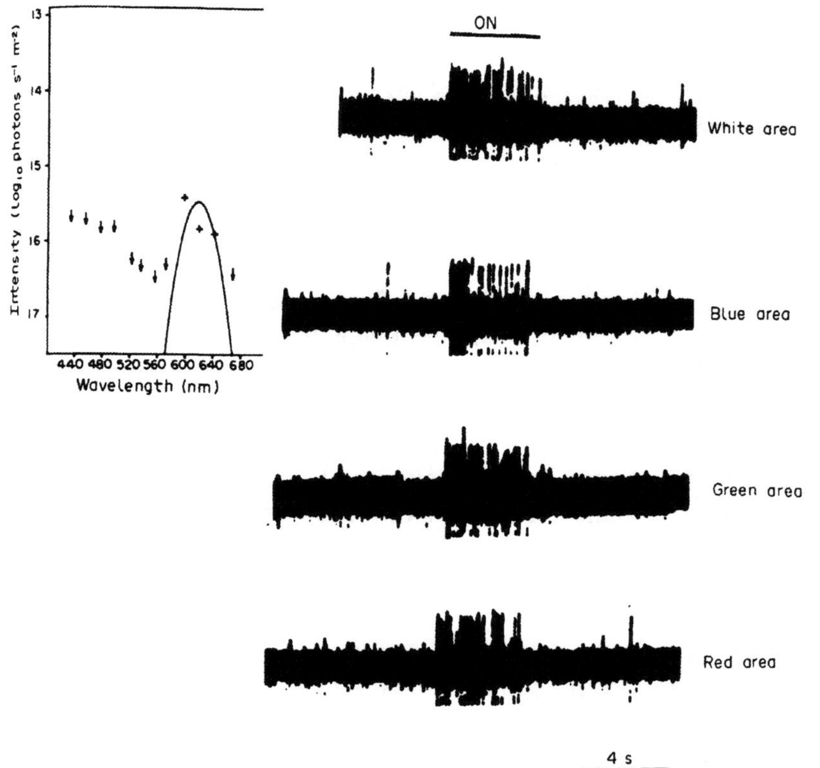

Fig. 4. The responses of a cell in monkey V1 to illumination of a display of many differently colored sheets of paper. Recordings of spike activity are shown in condensed time scale. Four different sheets, of the colors indicated on the *right,* were placed in turn under the receptive field, and different compositions of illuminating light were used so that each of those sheets reflected the same light. The cell was selective for light of long wavelengths, as shown by the spectral sensitivity plot in the *inset* (*downward arrows* indicate that no responses were obtained at the highest available intensities). Accordingly, an increased amount of long-wave light (compared to a standard composition) was necessary to evoke a response. However, once this condition was fulfilled, a response was evoked with any of the four papers under the receptive field, although each still appeared in its individual color, in spite of the different illuminations. The responses of this cell thus did not correlate with the color as seen by a human observer. (Reproduced, with permission, from Zeki 1983a)

display (which still appeared in their proper colors under these illuminations). Moreover, with different sequences of illumination (Zeki 1983b) such a cell could either be excited or silenced by stimulation with the same field under the same illumination. Consider, for example, a long-wave-sensitive cell. If the field was first illuminated by middle- plus short-wave light the cell would be excited when the long-wave light was added, but if the field was first illuminated by the long-wave light

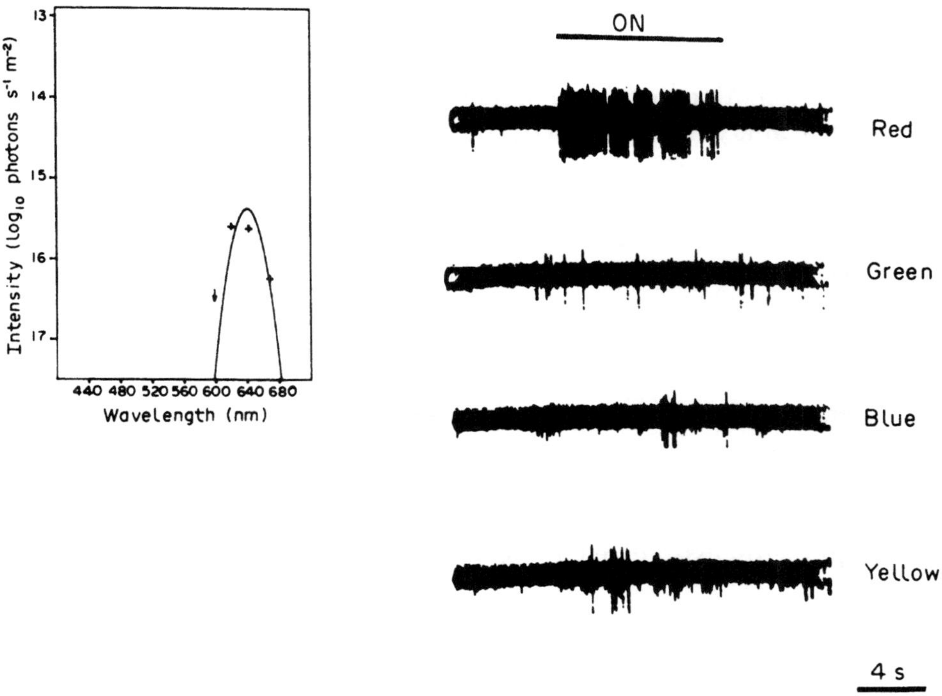

Fig. 5. The responses of a cell in V4 under similar conditions as those of Fig. 4. This cell again was selective for long-wave light, but illuminations that gave the standard composition of reflected light were used. Here, a response was obtained only when the red paper was centered under the receptive field, although each of the four papers reflected the same composition of light. Thus, the responses were said to correlate with perception of color. (Reproduced, with permission, from Zeki 1983a)

alone the cell would be silenced when the other two components were added. In either case, the field was finally illuminated by the same light, and appeared in the same color. By contrast, color-coded cells responded with excitation in both cases when the field was of the appropriate color.

Having found that responses of many wavelength selective cells did not correlate with the natural color of the stimulating field (i.e., its color when seen as part of the multicolored display), Zeki investigated whether their responses would correlate with its "void color", i.e., the color of a field when viewed in isolation, against a dark background; or whether the OFF responses of wavelength opponent cells would correlate with the color of the afterimage, either in the whole display, or in the void condition. In each case, examples of wavelength-dependent cells (in V1) were found that did not correlate with the observed colors (Zeki 1983a). An interesting perceptual result here was that the colors of afterimages (on a white screen), just as the directly viewed colors, did not have a simple relationship to wavelength; however, under various conditions of illumination, the color of the afterimage was complementary to the directly observed color. Another finding, perplexing at first sight, but in line with the theory, was that of color-coded cells (in V4) that did not respond to cut-outs of any color and responded poorly or not at all to monochromatic lights.

As Zeki points out, one should not read into his results that all cells of V4, but none of V1, are color coded. More extensive and quantitative studies are necessary to work out exactly the steps in processing between these areas. However, just knowing about the existence of those two types of responses in V1 and V4 opens a new perspective.

There is now also evidence from experiments with brain lesions and behavioral tests in monkey that V4 is involved in the kind of color perception discussed here (Wild et al. 1985).

5 Contour

As we have seen above, correlating neuronal responses with perception has led to very interesting results in the field of color vision. Perhaps the situation was particularly fortunate here. Zeki (1980) points out that "whereas it may be difficult to equate directly the responses of orientation selective cells with the perception of form, or of depth detecting cells with the perception of distance, there is little difficulty in equating the perception of colour with the response of the individual colour coded cells described here".

5.1 Contours and Neural Signals

There are ways, though, that may allow one to "equate" the responses of orientation selective cells, for example, with the perception of contour (Peterhans et al. 1982; von der Heydt et al. 1984). As in the case of color, one has to find a method of dissociating perceptual quality from physical stimulus property. In most physiological studies one could not distinguish whether the neural signals were related to contours or to light-dark boundaries, simply because in the stimuli that were used, contours always corresponded to light-dark boundaries. In fact the term "contour" was often used synonymously for "line" or "edge". (We reserve the term "contour" for a perception, and therefore avoid speaking of "perceived contours"; this is analogous to the use of the term "color" explained above.) Clearly, one cannot learn very much about mechanisms of form perception from just the responses to high-contrast rectangular stimuli in an otherwise empty field; detection and recognition of such an object is too easy. In many simple figures, contours can be perceived that are not due to luminance differences (see examples in Figs. 6 and 7).

Such contours are often called "illusory", although one can argue that contours frequently arise in a similar way under normal visual conditions, mostly when an object occludes structures in the background, and are then real, i.e., corresponding to object borders. The ability to perceive those illusory contours actually prevents illusions, because it allows detection of borders that would be undetectable by luminance gradients. One can avoid the problems with the term "illusory" by calling those contours "anomalous" (Lawson and Gulick 1967; cf. also Kanizsa 1979 for terminology). The comparison with the color experiments also shows how arbitrarily the term "illusion" is used. We can see an apple in a tree as red, even when it is illuminated by light scattered from the leaves and thus probably reflects more middle- than long-wave light, but do not speak of an illusion. Thus, what we call illusions obviously are not discrepancies between perception and physical reality, as one might think (in the case of the apple, the reflected light is the physical reality!), but contradictions between concepts, e.g., the visually derived concept of a contour and another concept that could be tactile, if, for example, the absence of a three-dimensional edge is recognized. For color, an alternative concept is not so readily available, since it requires spectral intensity measurements.

Figures 6 and 7 show neuronal responses recorded from the visual cortex of an alert rhesus monkey. She was trained to fixate a point on a display on which other stimuli were also presented. The fixation task required foveal visual acuity. Figure 6 shows responses of two cells, one recorded in area V1, the other in V2. Two kinds of stimuli were used, a light bar, and a pair of abutting gratings, as shown. In this grating stimulus, a vertical contour can be perceived although the stimulus consists only of horizontal lines. By presenting these two stimuli at various orientations it becomes very easy to distinguish whether a cortical neuron response correlates with perception of contour or with a local stimulus feature. As can be seen in Fig. 6A, the neuron in V1 responded when the bar was presented vertically, and when the gratings were shown with vertical lines, but not when the contour between the gratings was vertical. Thus, this kind of cell does not signal the line that we can see, although it signals orientation of

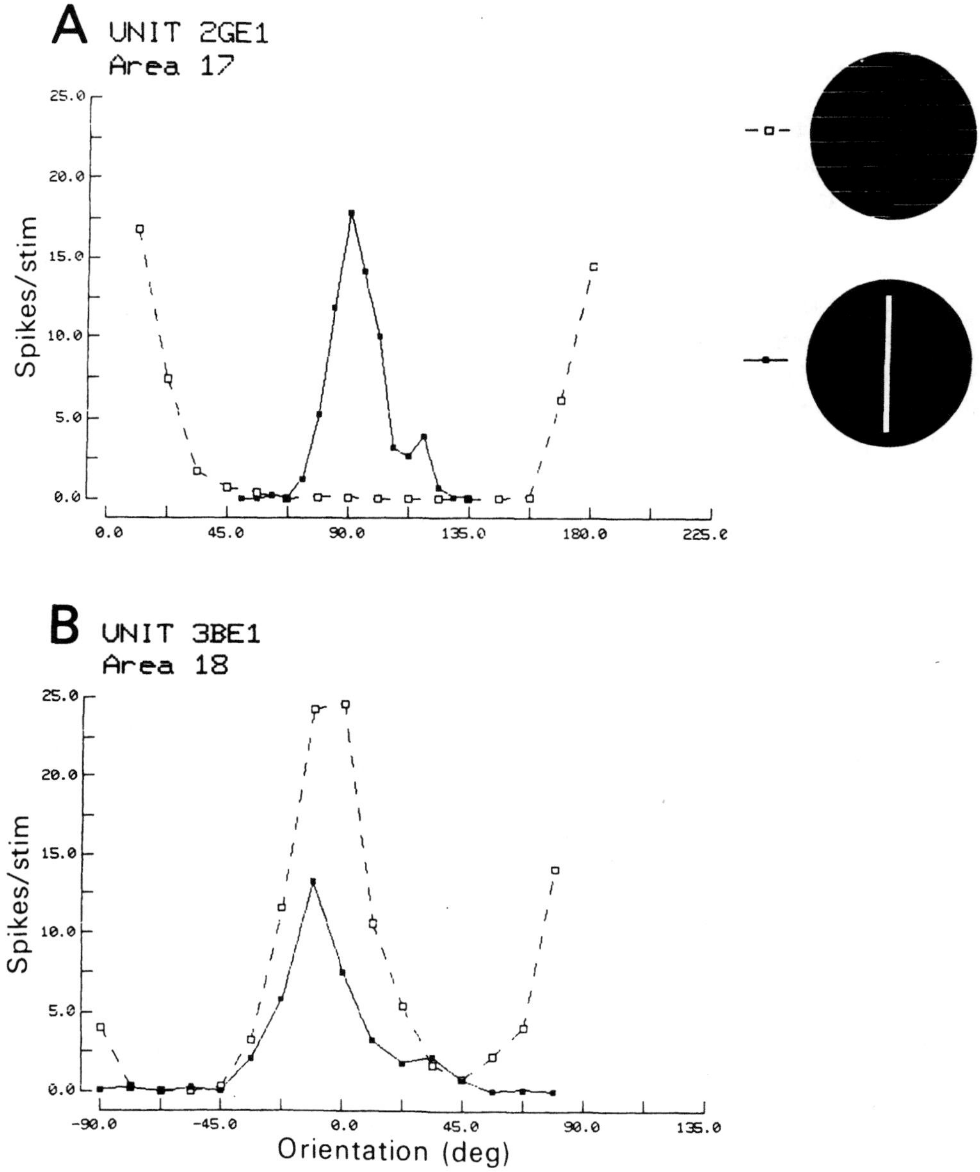

Fig. 6A, B. The responses of two cells in monkey visual cortex to light bars and to stimuli in which anomalous contours are perceived, as shown in the *insets.* Both stimuli were presented at various orientations, as indicated on the *abscissae*; the *angles* refer to the long axis of the bar and the line of discontinuity respectively. Both were oscillated perpendicularly to this orientation across the receptive fields. **A**, cell recorded in V1. The bar evoked the best response at vertical orientation, but the other stimulus did not produce any spike at this orientation, although a vertical contour was perceived. In order to evoke a response, it had to be rotated 90° so that the lines of the grating became vertical. these responses thus did not correlate with perception of contour. **B**, cell recorded in V2. To the bar, this cell responded best at about 10° from horizontal, and the other stimulus evoked an

luminance gradients with fidelity. The neuron shown in Fig. 6B also responded to a vertical bar, but responded also (in fact even better) to the grating stimulus when the contour between the gratings was vertical. It also responded when the lines of the gratings had vertical orientation. Thus, we can say that this cell responded whenever a vertical orientation could be perceived within its receptive field, even when this orientation was not given by a luminance gradient.

There are many ways to alter the perceptual quality of contour without changing the stimulus at the location where the contour is located. Figure 7 shows an example of a neural response, again recorded in V2, in such a condition. The stimulus here consisted of a pair of large dark rectangles that straddled a neuron's response field. The rectangles had notches on opposite sides that could be moved synchronously to and fro. In this case, a light bar is perceived moving in front of two black rectangles, bridging the gap between them. (Without movement, such a percept is usually obtained only when the "bar" is made wider; a narrow bar, as shown here, appears to pass behind the gap rather than in front.) The neuron whose responses are shown here was orientation selective and sensitive to the pattern; it responded to light bars (Fig. 7A) better than to edges or dark bars (not shown). It can be seen that it responded also, though more weakly, when the "illusory bar" moved across (Fig. 7B), although the actual stimulus was outside its response field (i.e., a single light bar that occupied exactly the position of one of the notches and moved exactly along the same path did not evoke a response). However, when the notches were closed by thin lines, the response was nearly abolished (Fig. 7C). In this case the illusory bar had also vanished.

When the stimuli of Figs. 6 and 7 were all tested in one cell (in V2), the same orientation preference was found for the solid bar, the illusory bar, and the border between gratings. Thus, responses were evoked whenever a contour of a certain orientation could be perceived at the center of the receptive field, with physically quite different stimuli. Several other observations have confirmed the assumption that the responses recorded in V2 were related to the perception of contour. One is the dependence on the number of lines in a stimulus as in Fig. 6. Just as the perceptual strength of the contour, the response was typically

even stronger response just when the anomalous contour had this orientation. One could thus activate this cell without any physical lines or edges at its preferred orientation. It also responded at an orientation 90° from this optimum, i.e., when the grating lines were oriented optimally. Thus, the responses of this cell seemed to correlate with the perception of lines or contours at a certain orientation. (Reproduced, with permission, from Baumgartner et al. 1984)

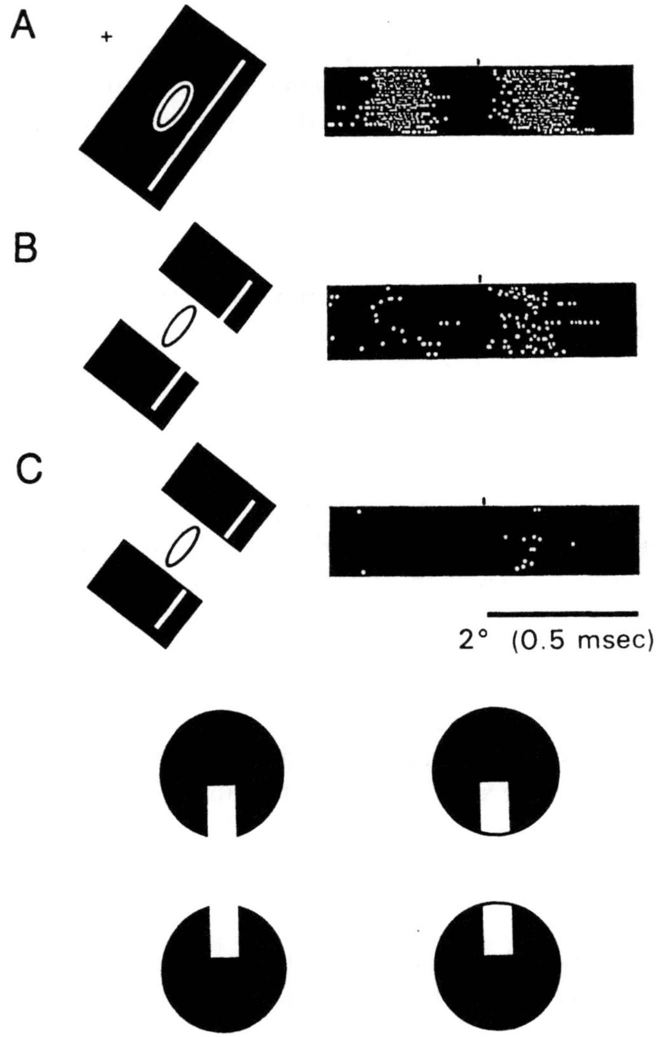

Fig. 7 A–C. Responses of a cell in monkey V2 related to perceptual completion. *Ellipses* represent the cell's response field, as mapped with a bar. A *cross* indicates the monkey's point of fixation. In **A**, a light bar was moved within a stationary black rectangle; the cell responded with 28 spikes per cycle on average. In **B**, a 2°-wide strip over the response field was blanked; this led to smaller, but still regular responses (4.6 spikes per cycle). In **C**, the two pieces of the bar were closed by thin lines (2 min arc); the responses were abolished (0.7 spikes per cycle). The perceptual illusion here depends in part on the movement. The *figures at the bottom* show a similar perceptual effect without movement. (Modified from von der Heydt et al. 1984).

zero with only one line and increased gradually up to many lines (von der Heydt et al. 1984). The tuning obtained with the anomalous contour stimulus (Fig. 6) was tuning to orientation and not direction of movement, because it could be obtained similarly with a stationary stimulus, and because, when grating lines and direction of movement were made to form an angle of 45°, instead of 90°, with the line of discontinuity, the response peak was determined by the orientation of this line and not by the direction of movement (E. Peterhans and R. von der Heydt, unpublished results). This was so in several sharply tuned cells. In others, the response peak tended to deviate toward the orientation perpendicular to the inducing lines (Baumgartner et al. 1984), which might have been an influence of movement direction, but could also be related to a perceptual deviation similar to the Zöllner illusion.

5.2 Stages of Processing

Responses that correlated with perception of contour were frequently found in V2 (about one-third of the cells tested) but not in V1 (von der Heydt et al. 1984). Until now, one meager example of a cell with a response to the abutting gratings that peaked at the orientation of the contour has been found in the deep layers of V1 (E. Peterhans and R. von der Heydt, unpublished results). It should be pointed out that absence of responses related to anomalous contours in V1 does not mean absence of responses to the stimuli that produce them. Clearly, all the information that leads to the perception of contours must be there in V1 (assuming that V2 gets all visual information via V1), as it must be present in the activity of cells in the retina and the dorsal lateral geniculate nucleus. The passage of a contour of the type of Fig. 6 can indeed be noticed in the responses of dLGN units and of various types of V1 cells. Particularly sharp responses were seen in end-stopped cells (hypercomplex cells, Hubel and Wiesel 1968) with receptive fields parallel to the lines, and thus perpendicular to the anomalous contour (E. Peterhans and R. von der Heydt, unpublished results). Von der Heydt et al. (1984) have therefore suggested that, in addition to the input from V1 simple or complex cells, end-stopped cells of V1 or V2 provide a second input to V2 cells which enables them to signals anomalous contours. These end-stopped cells should have orientations roughly orthogonal to the preferred orientations of their target cells in V2 (Fig. 8). Peterhans et al. (1984, 1986) have presented this model in more detail and reviewed the evidence for it. It explains the responses to the two types of anomalous contour stimuli shown in Fig. 6 and Fig. 7 by the same mechanism.

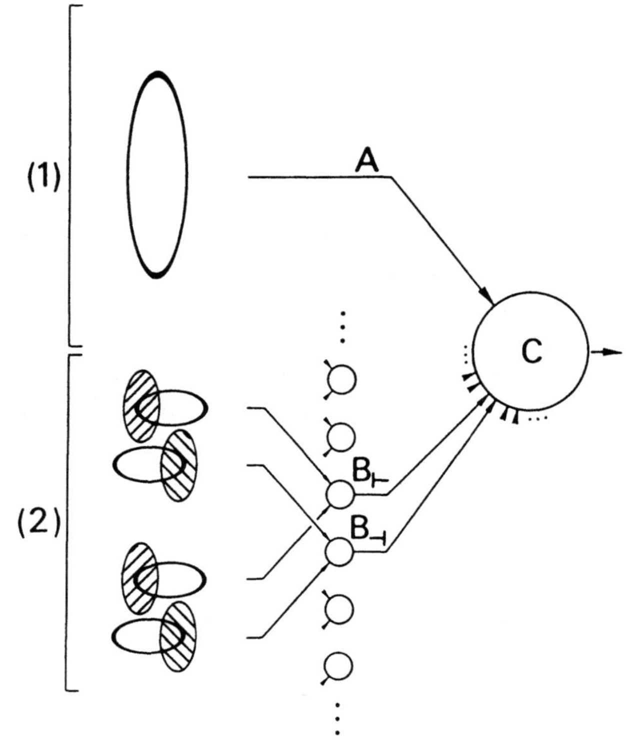

Fig. 8. A possible neural mechanism of contour perception. Cell **C**, assumed to signal contours of vertical orientation, receives excitatory inputs of two kinds: one from simple or complex cells of the same orientation (*1*), and a second input from asymmetrical end-stopped (hypercomplex) cells with orientations orthogonal to it (*2*). These cells are assumed to be excited by ends of lines and corners, but not by lines and edges extending across their inhibitory end zones (***hatched ellipses***). The units $B_{\vdash}$ and $B_{\dashv}$ are thought to receive inputs from two end-stopped cells, with receptive fields arranged on a vertical line, and to respond only when both inputs are excited. The receptive field groups (*1*) and (*2*) are thought to be centered on the same point in the retina and were drawn separately for clarity only. This scheme would allow the exploitation of luminance gradients (lines or edges of vertical orientation) as well as discontinuities (terminations of lines or edges) for the definition of a contour. (Reproduced, with permission, from Peterhans et al. 1986)

What is common to these rather different stimulus configurations is that end-stopped cells are excited, either by line ends (Fig. 6) or by corners (Fig. 7). The response reduction found when lines were added that intersected the ends of the illusory bar (Fig. 7C) is explained by the inhibition these lines cause in the end zones of those end-stopped cells. This model may also help to explain the perception of anomalous contours in a variety of conditions.

The Role of Hypercomplex Cells. Since the determination of object boundaries, particularly in images where objects partially occlude each other, is a key problem of vision, the neural mechanism of anomalous contour responses is of more general interest, particularly in the field of computer vision. According to current theories of edge detection, the gradients across boundaries are used for that task, an approach that fails when there are no detectable gradients, or when background structures interfere with their identification. The mechanism discussed here also exploits the discontinuity of background structures at the occluding contour. The additional information should result in more complete and valid representation of contour. We have suggested that this is the main function of end-stopped receptive fields (von der Heydt and Peterhans 1986). An exciting aspect of this new idea is that hypercomplex cells may not serve primarily as detectors for object features like corners, etc. (and I do not know of any alternative suggestions for their function), but in the analysis of occluding contour, where they are stimulated by characteristic features arising from the combination of foreground and background structures.

Redies et al. (1986) have demonstrated responses to stimuli of the type shown in Fig. 6 in cells of the dorsal lateral geniculate nucleus, and of areas 17 and 18, in the paralyzed, anesthetized cat. They scanned the receptive fields with the stimulus pattern, keeping the line of discontinuity parallel to the receptive field axes, and visualized the responses in two-dimensional displays of response frequency. For a linear filter-type receptive field such a display would correspond to the convolution of stimulus pattern and receptive field weighting function. In the displays of the responses of dLGN cells and of cortical complex cells in area 17 as well as area 18 the line of discontinuity (the anomalous contour) appeared marked by a frequency increase. The difference between the results in dLGN cells and cortical complex cells was that the dLGN cells marked each single line end of the stimulus, while the complex cells displayed a fused band of activity corresponding to the line of discontinuity. In displays of simple cells, the contour was, in general, not visible. There was thus a difference between simple and complex cells, but not between areas 17 and 18.

Redies et al. (1986) argue that the differences between their results and those of von der Heydt et al. (1984) are probably due to differences in neural organization between cat and monkey. Note, however, that the comparison is between anesthetized, paralyzed cats and alert, active monkeys. Comparing the responses to the abutting grating stimulus in the two studies, one finds indeed the difference that some complex cells in cat area 17 apparently responded well to the line of discontinuity (e.g., Fig. 1K of Redies et al. 1986) and to bars *at the same orientation,* whereas von der Heydt et al. (1984) found that, in V1, only nonoriented or poorly oriented cells responded to both stimuli at the same orientation. All oriented cells responded very weakly or not at all when the line of discontinuity was presented at their preferred orientation, in sharp contrast to V2. However, there are important differences in experimental design and scope. Redies et al. (1986) recorded responses at only one orientation of the anomalous contour stimulus. The cells might have responded to this stimulus at other orientations as well. Complex cells, but not simple cells in cat are known to respond well to moving textures (Hammond and MacKay 1977) and the line ends of the anomalous contour stimulus may produce similar responses. This interpretation would be supported by the observation that complex cells may "lose orientation" when signaling borders between textures (Nothdurft and Li 1984). If responses are to be related to perception of contour, I think evidence is needed that orientation is explicitly encoded, since it is an essential feature of contour perception. Even if tuning is found when the anomalous contour stimulus is presented at various orientations this could be tuning to the direction of movement rhather than orientation, as discussed above. Thus, one could have the impression that Redies et al. (1986) have found responses in cat area 17 that correlate with contour perception in the way that von der Heydt et al. (1984) did *not* find in monkey area 17, but this was not the case, and was not claimed. Nothdurft and Li (1984, 1985) found that responses in cat area 17 did not encode borders between textures of equal mean luminance.

An interesting observation of Redies et al. (1986) is that end-stopped (hypercomplex) cells in cat striate cortex responded when the end of a line, oriented and moved *at right angles* to the preferred orientation, entered, or left the receptive field. This is similar to the demonstration of Hubel and Wiesel (1965) that such cells often responded well to "light (or dark) tongues", except that in the experiment of Redies et al. the breadth of the tongue was reduced to the width of a line (0.1°); the stimulating edge, so to speak, had degenerated to the tip of a line. Redies et al. relate this observation to line-induced illusory changes in brightness and color (Ehrenstein 1941; van Tuijl 1975; Redies and Spillmann 1981). Peterhans and I have not seen such responses in the monkey, although we had expected to find them in V1.

5.3 Spatial Filter Theories

It is often assumed that illusory contours are physically present in spatially filtered images of the respective figures since this has been demonstrated for some of them (Ginsburg 1975; Becker and Knopp 1978). The stimulus of Fig. 6 has equal mean luminance on either side of the line of discontinuity. Indeed the mean taken along lines parallel to this line is constant all over the figure (disregarding the circular stimulus frame which was wide enough to have no effect on the neural responses). As a consequence, it has no Fourier components of the orientation of the contour, and nothing of that orientation can be brought out by filtering. Another figure with the same property has been designed by Parks and Pendergrass (1982), who established the perception of contours and verified the absence of those components, both in the optical Fourier transform and in filtered images. The objection was raised that the anomalous contour of Fig. 6 must be present in the Fourier phase spectrum. However, since the frequency components corresponding to the orientation of the contour have zero amplitudes, nothing can be encoded in their phases either. In fact, analytical calculation as well as numerical Fourier transforms of digitized images showed that the spectral power of the abutting grating stimulus is concentrated near the vertical frequency axis (for a figure with horizontal grating lines and vertical contour), whereas it is confined entirely to the vertical axis for horizontal gratings without the vertical contour; in neither case is there any power on the horizontal frequency axis where vertical bars have their Fourier components (R. von der Heydt, unpublished). Another possibility would be that nonlinear transduction of the image by the afferent system adds frequency components that are then detected by linear filtering at the cortical level. The tips of lines might be already highlighted at the level of the lateral geniculate responses, and cortical simple cells should be able to detect rows of such highlights by their elongated summation areas. Although this appears plausible, it was not borne out by the physiological experiments, where no responses corresponding to such contours were found in cortical simple cells (von der Heydt et al. 1984; Redies et al. 1986). In Fig. 7 we have seen results of another experiment speaking against the filter theory. The perception of the illusory bar as well as the neuronal responses related to it were affected drastically by the addition of line segments that only negligibly altered the light flux of the bar.

5.4 New Horizons

The anomalous contour figures are just examples of stimuli that could be used for dissociating perceptual and stimulus qualities. There are many other visual configurations which one could use for such purposes. Many have been designed for psychophysical studies, in attempts to identify mechanisms of perception. Kanizsa's book *Organization in Vision* (Kanizsa 1979) is a particularly rich thesaurus of such figures. Psychophysically, this task is often extremely difficult, and in many cases the experiments and demonstrations are ambiguous about the underlying perceptual mechanism. For the physiologist, I think, they are extremely valuable as means for distinguishing what is computed in a certain brain structure.

6 Relating Color and Contour

After the above discussions of color vision and contour mechanisms it would certainly be rewarding to find a section on the physiological relationship between colors and contours. However, I must pass here. It seems to me that nearly every study of cortical neuron responses, or none, could be cited. Since color is perceived as a qualitiy of surface, it is natural to relate the response of a cell to the surface seen by its receptive field. If, for example, the role of a concentric receptive field in color vision is judged by the criterion of whether it responds to a red spot on a green background (Livingstone and Hubel 1984a), or if signals of simple cells are related to color perception by the wavelength preference of the dominant subregion in their receptive field (Thorell et al. 1984; Vautin and Dow 1985), this seems to imply the assumption that the color of the area under the receptive field center is the concern of these signals.

On the other hand, it is clear from perceptual experiments that the color of a surface is determined at its borders. The Craik-O'Brian-Cornsweet brightness illusion is an example. Furthermore, by means of stabilized images one can demonstrate that large areas fill in with the color of the surround when the border-evoked neural activity ceases as a result of the stabilization, and small movements regenerating this activity bring back the original color (see Yarbus 1967 and Gerrits and Vendrik 1974). Similarly, retinal scotomata fill in immediately (Gerrits and Timmerman 1969); a normal observer can observe this phenomenon in his/her blind spots. In the extreme, if the border between red and green fields is stabilized, the whole field may even take on the two opponent colors simultaneously (Crane and Piantanida 1983). The Retinex algorithms also calculate the colors of multicolored displays ("Mondrians") on the basis of differences across the sharp borders of the stimulus fields. Accordingly, Livingstone and Hubel (1984a) postulate that broad-band and double-opponent cells play a role in color vision because these center-surround type receptive fields emphasize sharp borders. The double-opponent simple cells with two subregions of Thorell et al. (1984) would also suit this purpose. However, there is no hint so far as to how these border signals might be integrated to a color signal. A contour has two sides, and usually the colors on the two sides are different. How does the system know to which side those border-locked activities refer? How does it know which of the activities belong to the boundary of the same area, and how does it combine them to a single surface quality? How could one distinguish experimentally whether a neural signal refers to a surface or its boundary?

Does the perceptual phenomenon of filling-in imply that cortical activity spreads over the "empty" regions between borders, like water fills in the space between dykes? Spreading of cortical activity has been postulated in the theories of Gerrits and Vendrik (1970) and Grossberg and Mingolla (1985). Grossberg and Mingolla distinguish two contour processes, one that generates the "perceptual boundaries", something like a cartoon representation of the scene, and another that spreads brightness and colors until they hit those boundaries or decay. The boundaries gain visibility only by restricting the flow of brightness and color signals; neither of these dual processes alone is assumed to reach perception.

These "dyke theories" naturally appear wasteful to the single-unit physiologist. Some more serious criticisms, like the objection of naive realism, are discussed by Grossberg and Mingolla. It should be recognized that the postulated process results in a fairly direct representation of visual information by physical analogy; there is no symbolic description, which makes Marr's theory so interesting. On the other hand, Gerrits and Vendrik, as well as Grossberg and Mingolla, locate this process beyond area V1; the former, because filling-in occurs even with scotomata caused by lesion of V1, the latter, because the process requires as input the signals related to anomalous contours, which are found first in V2 (von der Heydt et al. 1984). Thus, after the filling-in, one might argue, not much of the visual cortex would be left for the rest of vision.

7 Conclusion

We have seen that knowledge about the receptive field structure of simple cells is still growing. In particular, quantitative descriptions have been made not only of their overt (excitatory) part, but also of their inhibitory portions that are often hidden from direct observation. To counter the impression that things are getting more and more complicated, I may point out that simple cells are now found to be more uniform as regards the number of subregions than previously thought. The interpretation of what these cells are doing in vision, I think, is still an open question. Simple cells are less simple than expected. I have pointed out that some difficulties arise from the fact that many studies have not distinguished between inhibitory and reciprocally excitatory effects, and have not paid attention to the implications of this distinction for the question of linearity. A general difficulty is that many studies of simple cells have obviously avoided tackling the question of their role in perception, perhaps for methodological or philosophical reasons. They were thus concerned not with the question "what are these cells doing", but "how are they doing it?" This policy was adopted when Hubel and Wiesel (1962) suggested that simple cells pool excitation of LGN signals with fields arranged along a line, that they sum (linearly) excitation and inhibition along their receptive field axis and across subregions, and that complex receptive fields are built up by pooling simple cell signals, suggestions that were based only on the knowledge of the cells' responses to flashing or moving rectangular shapes of light. Many subsequent studies have followed this route in trying to work out those receptive field concepts more quantitatively.

The studies of spatial filter properties show that one cannot easily deduce visual function from the receptive field concepts. Assuming that the simple cells constitute a stage of filtering, one has to accept that neither spatial frequency tuning nor extent of spatial sampling could be satisfactorily predicted from the previous analyses of their receptive fields. Assuming that simple cells are involved in edge detection as conjectured by the theory of Marr and Poggio (1979) and Marr and Hildreth (1980), nearly all questions are left open. The basic idea of this theory is that a comparison of the patterns of zero-crossings in a series of differently filtered versions of an image would allow the system to distinguish between various physical phenomena that may cause luminance variations. For example, by detecting a coincidence of zero-crossings in those filtered representations, it should be able to distinguish reflectance differences at occluding contours from luminance gradients caused by surface curvature or shadows. Marr and Hildreth (1980) and Marr and

Ullman (1981) thought that simple cells detect the zero-crossings and suggested a receptive field mechanism. Are the profiles and interaction plots, and the spatial frequency tunings of simple receptive fields compatible with this idea? The postulated coincidence detection would require cells whose receptive fields presumably would also appear simple. Whether simple or other cortical cells can distinguish between sharp and gradual luminance changes is hard to say, in spite of all our knowledge about receptive fields. It has never been tested. Other studies using new stimuli have also shown the weakness in predictive power of receptive field concepts, which is a handicap for theory building. From Hubel and Wiesel's simple-field concept one would have predicted responses to the line of discontinuity in the stimulus of Fig. 6, but no such responses were found in V1. From investigations of receptive fields in V2 using similar concepts, one would have predicted no difference from V1, but the observed responses to anomalous contour stimuli were quite different. In a related vein, could one tell from the studies of "end-inhibition" whether hypercomplex cells can signal the ends of lines in the pattern of Fig. 6, and if their signals would allow the system to localize the tips accurately enough so that the line of discontinuity could be detected? The conclusion must be that one needs to know the neuronal responses to a greater variety of stimuli than usually tested in receptive field studies. One might have thought that one could predict responses to posters of colored papers from the cells' wavelength tuning or from their spectral opponency, but responses of cells in V4 showed that this is not possible. Thus, in order to know how colored displays, illusory contour figures, or checkerboard patterns are processed in the cortex, one really needs to test them.

I have often taken up a position against the cortical Fourier analyzer theory, partly because of the evidence for nonlinear processes at early stages, but also because nobody seems to know how the cortex might proceed after the extraction of Fourier components. Nearly all the ideas about color, contour, depth, and form perception seem to follow from concepts of spatial, not frequency representations, such as Marr's primal sketch. I should emphasize again, however, that these ideas may also require linear filtering as a first step. One of the merits of the Fourier analyzer theory is certainly that it has drawn attention to the intermediate stages of processing between the image, as it is transferred to the cortex, and perception. Before, a more-or-less direct relationship between simple cell responses and perception of contour was often assumed, and color perception was related to LGN or retinal signals. The concept of spatial frequency filtering has spread the notion that simple cell signals only ambiguously represent stimulus features and may serve many purposes. It has made us familiar with the idea that something is computed from the image before perception takes place.

The new results relating cortical signals to color, contour, and movement perception strongly emphasize this point. A lot of processing obviously takes place in V1, but the result is a representation of visual information which still has little to do with perception. We do not perceive the output of V1, one might conclude. At higher levels, such as V2, V4, and MT, (some) signals do seem to correlate with perception, as far as we can see now. We do not yet understand how this is achieved. The attempts to relate neural signals to perception are attempts to connect physiological results with a large body of psychophysical and theoretical knowledge. Without referring to these attempts it would be unreasonable to demand that responses to a greater variety of simuli should be explored, given the enormous number of conceivably relevant stimuli, and regarding the studies, particularly in the last 10 years, that have investigated receptive field properties thoroughly and extensively, at a level that can hardly be surpassed. However, the concepts of perception indicate rather precisely what to look for. From theory and psychophysics we know, for example, that colors and contours are likely to be goals of cortical processing, and for what reasons, and we have at least suggestions as to what steps in processing might be involved. At the same time, the perceptions are still elementary so that identification with neural processes appears feasible. Thus, in the future, more physiological studies might exploit the concepts of perception and the ideas of the constructive approach.

The above discussion of some recent advances in visual cortical physiology may just give an impression of the amazing variety of interesting new ideas in this field. My selective review will, I hope, stimulate the interest to read the original reports (and occasionally to read between the lines) and could perhaps help to see some connections more celarly.

Acknowledgements. I thank Drs. Karen K. De Valois, Max R. Dürsteler, Walter H. Ehrenstein, and Esther Peterhans for critical comments on earlier drafts of the manuscript, and Dr. Vivien Kitteringham for correcting the English.

References

Akita M, Graham CH, Hsia Y (1964) Maintaining an absolute hue in the presence of of different background colors. Vision Res 4:539–556
Albrecht DG, De Valois RL (1981) Striate cortex responses to periodic patterns with and without the fundamental harmonics. J Physiol (Lond) 319:497–514
Albrecht DG, Hamilton DB (1982) Striate cortex of monkey and cat: contrast response function. J Neurophysiol 48:217–237
Albrecht DG, De Valois RL, Thorell LG (1980) Visual cortical neurons: are bars or gratings the optimal stimuli? 207:88–90
Albrecht DG, Farrer SB, Hamilton DB (1984) Spatial contrast adaptation characteristics of neurones recorded in the cat's visual cortex. J Physiol (Lond) 347: 713–739
Allman J, Miezin F, McGuinness E (1985a) Direction- and velocity-specific responses from beyond the classical receptive field in the middle temporal visual area (MT). Perception 14:105–126
Allman J, Miezin F, McGuinness E (1985b) Stimulus specific responses from beyond the classical receptive field: neurophysiological mechanisms for local-global comparisons in visual neurons. Annu Rev Neurosci 8:407–430
Andersen VO, Guld C, Sjö O (1983) Colour processing in prestriate cortex of vervet monkey. In: Mollon JD, Sharpe LT (eds) Colour vision. Physiology and psychophysics. Academic London, pp 297–301
Andrews BW, Pollen DA (1979) Relationship between spatial frequency selectivity and receptive field profile of simple cells. J Physiol (Lond) 287:163–176
Badcock DR, Derrington AM (1985) Detecting the displacement of periodic patterns. Vision Res 25:1253–1258
Barlow HB (1978) The efficiency of detecting changes of density in random dot patterns. Vision Res 18:637–650
Bauer R (1982) A high probability of an orientation shift between layers 4 and 5 in central parts of the cat striate cortex. Exp Brain Res 48:245–255
Bauer R, Dow BM, Snyder AZ, Vautin R (1983) Orientation shift between upper and lower layers in monkey visual cortex. Exp Brain Res 50:133–145
Baumgartner G (1960) Indirekte Größenbestimmung der rezeptiven Felder der Retina beim Menschen mittels der Hermannschen Gittertäuschung. Pflügers Arch 272:21–22
Baumgartner G, Hakas P (1962) Die Neurophysiologie des simultanen Helligkeitskontrastes. Reziproke Reaktionen antagonistischer Neuronengruppen des visuellen Systems. Pflügers Arch 274:489–510
Baumgartner G, von der Heydt R, Peterhans E (1984) Anomalous contours: a tool in studying the neurophysiology of vision. Exp Brain Res [Suppl] 9:413–419
Becker MF, Knopp J (1978) Processing of visual illusions in the frequency and spatial domains. Percept Psychophys 23:521–526
Berardi N, Bisti S, Cattaneo A, Fiorentini A, Maffei L (1982) Correlation between the preferred orientation and spatial frequency of neurones in visual areas 17 and 18 of the cat. J Physiol (Lond) 323:603–618
Bishop PO (1984) Processing of visual information within the retinostriate system. In: Darian-Smith I (ed) The nervous system. Am Physiol Soc, Bethesda, pp 341–424. (Handbook of Physiology, vol 3 sec 1)
Bishop PO, Henry GH (1972) Striate neurons: receptive field concepts. Invest Ophthalmol Vis Sci 11:346–354
Bishop PO, Henry GH, Smith CJ (1971) Binocular interaction fields of single units in the cat striate cortex. J Physiol (Lond) 216:39–68
Bishop PO, Combs JS, Henry GH (1973) Receptive fields of simple cells in the cat striate cortex. J Physiol (Lond) 231:31–60

Blasdel GG, Fitzpatrick O (1984) Physiological organization of layer 4 in macaque striate cortex. Neuroscience 4:880–895
Bolz J, Gilbert CD (1986) Generation of end-inhibition in the visual cortex via interlaminar connections. Nature 320:362–365
Bruce CJ, Isley MR, Shinkman PG (1981) Visual experience and development of interocular disparity in visual cortex. J Neurophysiol 46:215–228
Camarda RM, Bishop PO, Peterhans E (1979) Inhibitory sidebands in the receptive fields of cells in the simple family in cat striate cortex. Proc Austral Physiol Pharmacol Soc 10:11P
Camarda RM, Peterhans E, Bishop PO (1985a) Spatial organization of subregions in receptive fields of simple cells in cat striate cortex as revealed by stationary flashing bars and moving edges. Exp Brain Rex 60:136–150
Camarda RM, Peterhans E, Bishop PO (1985b) Simple cells in striate cortex: responses to stationary flashing and to moving light bars. Exp Brain Res 60:151–158
Campbell FW, Maffei L (1979) Stopped visual motion. Nature 278:192
Campbell FW, Robson JG (1968) Application of Fourier analysis to the visibility of gratings. J Physiol (Lond) 197:551–566
Campbell FW, Howell ER, Johnstone JR (1978) A comparison of threshold and suprathreshold appearance of gratings with components in the low and high spatial frequency range. J Physiol (Lond) 284:193–201
Carlson CR, Moeller JR, Anderson CH (1984) Visual illusions without low spatial frequencies. Vision Res 24:1407–1413
Carlson M, Hubel DH, Wiesel TN (1986) Effects of monocular exposure to oriented lines on monkey striate cortex. Exp Brain Res 25:71–82
Cogan AI (1979) The relationship between the apparent vertical and the vertical horopter. Vision Res 19:655–665
Crane HD, Piantanida TP (1983) On seeing reddish green and yellowish blue. Science 221:1078–1080
Creutzfeldt O, Ito M (1968) Functional synaptic organization of primary visual cortex neurones in the cat. Exp Brain Res 6:324–352
Creutzfeldt OD, Nothdurft HC (1978) Representation of complex visual stimuli in the brain. Naturwissenschaften 65:307–318
Daugman JG (1980) Two-dimensional spectral analysis of cortical receptive field profiles. Vision Res 20:847–856
Daugman JG (1983) Six formal properties of two-dimensional anisotropic visual filters: structural principles and frequency/orientation selectivity. IEEE Trans SMC 13:882–887
Daugman JG (1984) Spatial visual channels in the Fourier plane. Vision Res 24: 891–910
Daw NW (1968) Colour-coded ganglion cells in the goldfish retina: extension of their receptive fields by means of new stimuli. J Physiol (Lond) 197:567–592
Daw NW (1984) The psychology and physiology of color vision. Trends Neurosci 7:330–335
Dean AF (1981) The relationship between response amplitude and contrast for cat striate neurones. J Physiol (Lond) 318:413–427
Dean AF, Tolhurst DJ (1983) On the distinctness of simple and complex cells in the visual cortex of the cat. J Physiol (Lond) 344:305–325
Dean AF, Tolhurst DJ, Walker NS (1982) Non-linear temporal summation by simple cells in cat striate cortex demonstrated by failure of superposition. Exp Brain Res 45:456–458
De Monasterio FM, Schein SJ (1982) Spectral bandwidths of color-opponent cells of geniculocortical pathway of macaque monkeys. J Neurophysiol 47:214–224
Derrington AM, Krauskopf J, Lennie P (1984) Chromatic mechanisms in lateral geniculate nucleus of macaque. J Physiol (Lond) 357:241–265

Desimone R, Schein SJ, Moran J, Ungerleider LG (1985) Contour, color and shape analysis beyond the striate cortex. Vision Res 25:441–452
De Valois RL, De Valois KK (1980) Spatial vision. Annu Rev Psychol 31:309–341
De Valois RL, Pease PL (1971) Contours and contrast: responses of macaque lateral geniculate cells to color and luminance figures. Science 171:694–696
De Valois KK, Tootell RBH (1983) Spatial-frequency-specific inhibition in cat striate cortex cells. J Physiol (Lond) 336:359–376
De Valois RL, Albrecht DG, Thorell L (1978) Cortical cells: bar detectors or spatial frequency filters? In: Cool SJ, Smith EL (eds) Frontiers in visual science. Springer, Berlin Heidelberg New York, pp 544–556
De Valois KK, De Valois RL, Yund EW (1979) Responses of striate cortex cells to grating and checkerboard patterns. J Physiol (Lond) 291:483–505
De Valois RL, Albrecht DG, Thorell LG (1982) Spatial frequency selectivity of cells in macaque visual cortex. Vision Res 22:545–559
De Valois RL, Thorell LG, Albrecht DG (1985) Periodicity of striate-cortex-cell receptive fields. J Opt Soc Am 2:1115–1123
Dow BM (1974) Functional classes of cells and their laminar distribution in monkey visual cortex. J Neurophysiol 37:927–946
Dürsteler MR, von der Heydt R (1983) Plasticity in the binocular correspondence of striate cortical receptive fields in kittens. J Physiol (Lond) 345:87–105
Ehrenstein W (1941) Ueber Abwandlungen der L. Hermannschen Helligkeitserscheinung. 150:83–91
Emerson RC, Gerstein GL (1977a) Simple striate neurons in the cat. I. Comparison of responses to moving and stationary stimuli. J Neurophysiol 40:119–135
Emerson RC, Gerstein GL (1977b) Simple striate neurons in the cat. II. Mechanisms underlying directional asymmetry and directional selectivity. J Neurophysiol 40:136–155
Foster KH, Gaska JP, Nagler M, Pollen DA (1985) Spatial and temporal frequency selectivity of neurones in visual cortical areas V1 and V2 of the macaque monkey. J Physiol (Lond) 365:331–363
Gerrits HJM, Timmerman GJMEN (1969) The filling-in process in patients with retinal scotoma. Vision Res 9:439–442
Gerrits HJM, Vendrick AJ (1970) Simultaneous contrast, filling-in process and information processing in man's visual system. Exp Brain Res 11:411–430
Gerrits HJM, Vendrik AJH (1974) The influence of stimulus movements on perception in parafoveal stabilized vision. Vision Res 14:175–180
Ginsburg AP (1975) Is the illusory triangle physical or imaginary? Nature (Lond) 257:219–220
Ginsburg AP (1982) On a filter approach to understanding the perception of visual form. In: Albrecht DG (ed) Recognition of pattern and form. Springer, Berlin Heidelberg New York, pp 175–192
Glezer VD, Ivanoff VA, Tscherbach TA (1973) Investigation of complex and hypercomplex receptive fields of visual cortex of the cat as spatial frequency filters. Vision Res 13:1875–1904
Glezer VD, Tsherbach TA, Gauselman VE, Bondaroko VM (1980) Linear and nonlinear properties of simple and complex receptive fields in area 17 of the cat visual cortex. Biol Cybern 37:195–208
Glezer VD, Tsherbach TA, Gauselman VE, Bondarko VM (1982) Spatio-temporal organisation of receptive fields of the cat striate cortex. Biol Cybern 43:35–49
Gonzales RC, Wintz P (1977) Digital image processing. Adison-Wesley, Reading
Gouras P (1974) Opponent-colour cells in different layers of foveal striate cortex. J Physiol (Lond) 238:583–602
Gouras P, Krüger J (1979) Responses of cells in foveal visual cortex of the monkey to pure color contrast. J Neurophysiol 42:850–860

Grosof DH, Skottun BC, De Valois RL (1985) Linearity of cat cortical cells: responses to 2-d stimuli. Invest Ophthalmol Vsi Sci 26:265

Grossberg S, Mingolla E (1985) Neural dynamics of form perception: boundary completion, illusory figures, and neon color spreading. Psychol Rev 92:173–211

Hagerty CM, Lees FC, Tieman SB, Hirsch HVB (1982) Principal components analysis of cells in cat visual cortex. Brain Res 251:45–53

Hamilton DB, Albrecht DG, Geisler WS (1985) Receptive field organization of visual cortical neurons in monkey and cat: the phase and amplitude transfer functions. Invest Ophthalmol Vis Sci 26:265

Hammond P, MacKay DM (1977) Differential responsiveness of simple and complex cells in cat striate cortex to visual texture. Exp Brain Res 30:275–296

Hammond P, MacKay DM (1981) Modulatory influences of moving textured backgrounds on responsiveness of simple cells in feline striate cortex. J Physiol (Lond) 319:431–442

Hammond P, MacKay DM (1983) Influence of luminance gradient reversal on simple cells in feline striate cortex. J Physiol (Lond) 337:69–87

Hammond P, MacKay DM (1985) Influence of luminance gradient reversal on complex cells in feline striate cortex. J Physiol (Lond) 359:315–329

Hänny P, von der Heydt R (1982) Effect of horizontal-plane environment on the development of binocular receptive fields of cells in cat visual cortex. J Physiol (Lond) 329:75–92

Hawken MJ, Parker AJ (1984) Contrast sensitivity and orientation selectivity in lamina IV of striate cortex of Old World monkeys. Exp Brain Res 54:367–372

Heggelund P (1981a) Receptive field organisation of simple cells in cat striate cortex. Exp Brain Res 42:89–98

Heggelund P (1981b) Receptive field organisation of complex cells in cat striate cortex. Exp Brain Res 42:99–107

Heggelund P (1986a) Quantitative studies of the discharge fields of single cells in cat striate cortex. J Physiol (Lond) 373:277–292

Heggelund P (1986b) Quantitative studies of enhancement and suppression zones in the receptive field of simple cells in cat striate cortex. J Physiol (Lond) 373:293–310

Heggelund P, Krekling S, Skottun BC (1984) Spatial summation in subregions of simple-cell receptive fields in cat striate cortex as a function of slit length. J Physiol (Lond) 352:327–338

Helmholtz H (1864) Über den Horopter. Albrecht v Graefe's Arch Ophthal 10:1–60

Henning GB, Hertz BG, Broadbent DE (1975) Some experiments bearing on the hypothesis that the visual system analyses spatial patterns in independent bands of spatial frequency. Vision Res 15:887–898

Henry GH, Bishop PO, Coombs JS (1969) Inhibitory and subliminal excitatory receptive fields of simple units in cat striate cortex. Vision Res 9:1289–1296

Hering E (1878) Zur Lehre vom Lichtsinne. Sechs Mittheilungen an die kaiserl. Akademie der Wissenschaften in Wien. Carl Gerold's Sohn, Vienna

Hering E (1964) Qutlines of a theory of the light sense. Translated by LM Hurvich and D Jameson. Harvard University Press, Cambridge

Horn BKP (1974) Determining lightnesses from an image. Comput Graphics Image Process 3:277–299

Hubel DH, Wiesel TN (1962) Receptive fields, binocular interaction and functional architecture in the cat's visual cortex. J Physiol (Lond) 160:106–154

Hubel DH, Wiesel TN (1965) Receptive fields and functional architecture in two nonstriate visual areas (18 and 19) of the cat. J Neurophysiol 28:229–289

Hubel DH, Wiesel TN (1968) Receptive fields and functional architecture of monkey striate cortex. J Physiol (Lond) 195:215–243

Hubel DH, Wiesel TN (1974) Sequence regularity and geometry of orientation columns in the monkey striate cortex. J Comp Neurol 158:267–294

Jung R (1972) Neurophysiological and psychophysical correlates in vision research. In: Karczmar AG, Eccles Sir JC (eds) Brain and human behaviour. Springer, Berlin Heidelberg New York, pp 209–258
Jung R (1973) Visual perception and neurophysiology. In: Jung R (ed) Central visual information. Springer, Berlin Heidelberg New York, pp 1–152 (Handbook of sensory physiology, vol 7/3A)
Kanizsa G (1979) Organization in vision. Essays on gestalt perception. Praeger, New York
Kato H, Bishop PO, Orban GA (1978) Hypercomplex and simple/complex cell classifications in cat striate cortex. J Neurophysiol 41:1071–1095
Kelly DH (1976) Pattern detection and the two-dimensional Fourier transform: flickering checkerboards and chromatic mechanisms. Vision Res 16:277–287
Kelly DH, Burbeck CA (1984) Critical problems in spatial vision. CRC Crit Rev Biomed Engin 10:125–177
Koch C, Poggio T (1983) A theoretical analysis of electrical properties of spines. Proc R Soc Lond Biol 218:455–477
Krüger J, Bach M (1982) Independent systems of orientation columns in upper and lower layers of monkey visual cortex. Neurosci Lett 31:225–230
Krüger J, Gouras P (1980) Spectral sensitivity of cells and its dependence on slit length in monkey visual cortex. J Neurophysiol 43:1055–1069
Kulikowski JJ, Bishop PO (1981a) Linear analysis of the responses of simple cells in the cat visual cortex. Exp Brain Res 44:386–400
Kulikowski JJ, Bishop PO (1981b) Fourier analysis and spatial representation in the visual cortex. Experientia 37:160–163
Kulikowski JJ, Bishop PO (1982) Silent periodic cells in the cat striate cortex. Vision Res 22:191–200
Kulikowski JJ, Bishop PO, Kato H (1981) Spatial arrangements of responses by cells in the cat visual cortex to light and dark bars and edges. Exp Brain Res 44:371–385
Kulikowski JJ, Marcelja S, Bishop PO (1982) Theory of spatial position and spatial frequency relations in the receptive fields of simple cells in the visual cortex. Biol Cybern 43:187–198
Land EH (1959a) Color vision and the natural image. Part I. Proc Natl Acad Sci USA 45:115–129
Land EH (1959b) Color vision and the natural image. Part II. Proc Natl Acad Sci USA 45:636–644
Land EH (1983) Recent advances in Retinex theory and some implications for cortical computations: color vision and the natural image. Proc Natl Acad Sci USA 80:5161–5169
Land EH (1986) Recent advances in Retinex theory. Vision Res 26:7–22
Land EH, McCann JJ (1971) Lightness and Retinex theory. J Opt Soc Am 61: 1–11
Land EH, Hubel DH, Livingstone MS, Perry SH, Burns M (1983) Colour-generating interactions across the corpus callosum. Nature (Lond) 303:616–618
Lawson RB, Gulick WL (1967) Stereopsis and anomalous contour. Vision Res 7:271–297
Levine J, Spillmann L, Wolf E (1980) Saturation enhancement in colored Hermann grids varying only in chroma. Vision Res 20:307–313
Livingstone MS, Hubel DH (1984a) Anatomy and physiology of a color system in the primate visual cortex. J Neurosci 4:309–356
Livingstone MS, Hubel DH (1984b) Specificity of intrinsic connections in primate primary visual cortex. J Neurosci 4:2830–2835
MacKay DM (1981) Strife over visual cortical function. Nature (Lond) 289:117–118
MacKay DM (1985) The significance of 'feature sensitivity'. In: Rose D, Dobson VG (eds) Models of the visual cortex. Wiley, Chichester, pp 47–53

Maffei L, Fiorentini A (1973) The visual cortex as a spatial frequency analyser. Vision Res 13:1255–1267
Maffei L, Fiorentini A (1976) The unresponsive regions of visual cortical receptive fields. Vision Res 16:1131–1139
Maffei L, Fiorentini A, Bisti S (1973) Neural correlate of perceptual adaptation to gratings. Science 182:1036–1038
Maffei L, Morrone C, Pirchio M, Sandini G (1979) Responses of visual cortical cells to periodic and non-periodic stimuli. J Physiol (Lond) 296:27–47
Marcelja S (1980) Mathematical description of the responses of simple cortical cells. J Opt Soc Am 70:1297–1300
Marr D (1982) Vision. A computational investigation into the human representation and processing of visual information. Freeman, San Francisco, p 75
Marr D, Hildreth E (1980) Theory of edge detection. Proc R Soc Lond Biol 207: 187–217
Marr D, Poggio T, (1979) A computational theory of human stereo vision. Proc R Soc Lond Biol 204:301–328
Marr D, Ullman S (1981) Directional selectivity and its use in early visual processing. Proc R Soc Lond Biol 211:151–180
Maske R, Yamane S, Bishop PO (1985) Simple and B-cells in cat striate cortex: complementarity of responses to moving light and dark bars. J Neurophysiol 53:670–685
McCann J, McKee P, Taylor H (1976) A comparison between theoretical predictions and observer responses to the "Color Mondrian" experiments. Vision Res 16: 445–458
McCarter A (1979) Chromatic induction effects in the Hermann grid illusion. Perception 8:105–114
Michael CR (1978a) Color vision mechanisms in monkey striate cortex: dual-opponent cells with concentric receptive fields. J Neurophysiol 41:572–588
Michael CR (1978b) Color vision mechanisms in monkey striate cortex: simple cells with dual opponent-color receptive fields. J Neurophysiol 41:1233–1249
Michael CR (1978c) Color-sensitive complex cells in monkey striate cortex. J Neurophysiol 41:1250–1266
Michael R (1979) Color-sensitive hypercomplex cells in monkey striate cortex. J Neurophysiol 42:726–744
Michael CR (1985) Laminar segregation of color cells in the monkey's striate cortex. Vision Res 25:415–423
Mimura K (1986) Development of visual pattern discrimination in the fly depends on light experience. Science 232:83–85
Morrone MC, Burr DC, Maffei L (1982) Functional implications of cross-orientation inhibition of cortical visual cells. I. Neurophysiological evidence. Proc R Soc Lond Biol 216:335–354
Movshon JA, Lennie P (1979) Pattern selective adaptation in visual cortical neurones. Nature (Lond) 278:850–851
Movshon JA, Van Sluyters RC (1981) Visual neural development. Annu Rev Psychol 32:477–522
Movshon JA, Thompson ID, Tolhurst DJ (1978a) Spatial summation in the receptive fields of simple cells in the cat's striate cortex. J Physiol (Lond) 283:53–77
Movshon JA, Thompson ID, Tolhurst DJ (1978b) Receptive field organization of complex cells in the cat's striate cortex. J Physiol (Lond) 283:79–99
Movshon JA, Thompson ID, Tolhurst DJ (1978c) Spatial and temporal contrast sensitivity of neurones in areas 17 and 18 of the cat's visual cortex. J Physiol (Lond) 283:101–120
Movshon JA, Adelson EH, Gizzi MS, Newsome WT (1985) The analysis of moving patterns. Exp Brain Res [Suppl] 11:117–151

Mullikin WH, Jones JP, Palmer LA (1984a) Receptive field properties and laminar distribution of X-like and Y-like simple cells in cat area 17. J Neurophysiol 52:350–371

Mullikin WH, Jones JP, Palmer LA (1984b) Periodic simple cells in cat area 17. J Neurophysiol 52:372–387

Nachmias J, Sansbury R, Vassilev A, Weber A (1973) Adaptation to square-wave gratings: in search of the elusive third harmonic. Vision Res 13:1335–1342

Nelson JI, Forst BJ (1978) Orientation-selective inhibition from beyond the classic visual receptive field. Brain Res 139:359–365

Nelson JI, Frost BJ (1985) Intracortical facilitation among co-oriented, co-axially aliged simple cells in cat striate cortex. Exp Brain Res 61:54–61

Nothdurft HC, Li CY (1984) Representation of spatial details in textured patterns by cells of the cat striate cortex. Exp Brain Res 57:9–21

Nothdurft HC, Li CY (1985) Texture discrimination: representation of orientation and luminance differences in cells of the cat striate cortex. Vision Res 25: 99–113

Oehler R, Spillmann L (1981) Illusory colour changes in Hermann grids varying only in hue. Vision Res 21:527–541

Ohzawa I, Sclar G, Freeman RD (1982) Contrast gain control in the cat visual cortex. Nature (Lond) 298:266–268

Oyama T, Hsia Y (1966) Compensatory hue shift in simultaneous color contrast as a function of separation between inducing and test fields. J Exp Psychol 71:405–413

Palmer LA, Davis TL (1981a) Receptive-field structure in cat striate cortex. J Neurophysiol 46:260–276

Palmer LA, Davis TL (1981b) Comparison of responses to moving and stationary stimuli in cat striate cortex. J Neurophysiol 46:277–295

Parks E, Pendergrass L (1982) On the filtered-compounds approach to illusory contours. Percept Psychophys 32:491–493

Pease PL (1978) On color Mach bands. Vision Res 18:751–755

Peterhans E, von der Heydt R, Baumgartner G (1982) Cortical neuron responses and 'illusory' contours. Perception 11:A19

Peterhans E, von der Heydt R, Baumgartner G (1984) Illusory contour stimuli reveal stages in visual cortical processing. Perception 13:A16–A17

Peterhans E, von der Heydt R, Baumgartner G (1986) Neuronal responses to illusory contour stimuli reveal stages of visual cortical processing. In: Pettigrew JD, Sanderson KJ, Levick WR (eds) Visual neuroscience. Cambridge University Press, Cambridge, pp 343–351

Petrov AP, Pigarev IN, Zenkin GM (1980) Some evidence against Fourier analysis as a function of the receptive fields in cat's striate cortex. Vision Res 20:1023–1026

Pettigrew JD, Nikara T, Bishop PO (1968) Responses to moving slits by single units in cat striate cortex. Exp Brain Res 6:373–390

Poggio GF (1972) Spatial properties of neurons in striate cortex of unanaesthetized macaque monkey. Invest Ophthalmol Vis Sci 11:368–377

Poggio GF, Baker FH, Mansfield RJW, Sillito A, Grigg P (1975) Spatial and chromatic properties of neurons subserving foveal and parafoveal vision in rhesus monkeys. Brain Res 100:25–59

Poggio GF, Doty RW, Talbot WH (1977) Foveal striate cortex of behaving monkey: single neuron responses to square-wave gratings during fixation of gaze. J Neurophysiol 40:1369–1391

Poggio T (1981) Marr's computational approach to vision. Trends Neurosci 4: 258–262

Poggio T, Koch C (1985) Ill-posed problems in early vision: from computational theory to analogue networks. Proc R Soc Lond Biol 226:303–323

Poggio T, Torre V (1978) A new approach to synaptic interactions. In: Heim R, Palm G (eds) Approaches to complex systems. Springer, Berlin Heidelberg New York, pp 89–115
Pollen DA, Ronner SF (1982) Spatial computation performed by simple and complex cells in the visual cortex of the cat. Vision Res 22:101–118
Pollen DA, Ronner SF (1983) Visual cortical neurons as localized spatial frequency filters. IEEE Trans SMC 13:907–916
Pollen DA, Lee JR, Taylor JH (1971) How does the striate cortex begin the reconstruction of the visual world? Science 173:74–77
Pöppel E (1986) Long-range colour generating interactions across the retina. Nature (Lond) 320:523–525
Rauschecker JP, Singer W (1981) The effects of early visual experience on the cat's visual cortex and their possible explanation by Hebb synapses. J Physiol (Lond) 310:215–239
Redies C, Spillmann L (1981) The neon-color effect in the Ehrenstein illusion. Perception 10:667–681
Redies C, Crook JM, Creutzfeldt OD (1986) Neuronal responses to borders with and without luminance gradients in cat visual cortex and dorsal lateral geniculate nucleus. Exp Brain Res 61:469–481
Reichardt W (1961) Autocorrelation, a principle for evaluation of sensory information by the central nervous system. In: Rosenblith WA (ed) Principles of sensory communication. Wiley, New York, pp 303–317
Reichardt W (1986) Processing of optical information by the visual system of the fly. Vision Res 26:113–126
Robson JG (1980) Neural images: the physiological basis of spatial vision. In: Harris C (ed) Visual coding and adaptability. Erlbaum, Hillsdale, pp 177–214
Sagi D, Julesz B (1985) "Where" and "what" in vision. Science 228:1217–1219
Schein SJ, Marrocco RT, De Monasterio FM (1982) Is there a high concentration of color-selective cells in area V4 of monkey visual cortex? J Neurophysiol 47:193–213
Schiller PH (1982) Central connections of the retinal ON and OFF pathways. Nature 297:580–583
Schiller PH, Finlay BL, Volman SF (1976a) Quantitative studies of single-cell properties in monkey striate cortex. I. Spatiotemporal organization of receptive fields. J Neurophysiol 39:1288–1319
Schiller PH, Finlay BL, Volman SF (1976b) Quantitative studies of single-cell properties in monkey striate cortex III. Spatial frequency. J Neurophysiol 39:1334–1351
Schumer RA, Movshon JA (1984) Length-summation in simple cells of cat striate cortex. Vision Res 24:565–571
Shapley R, Lennie P (1985) Spatial frequency analysis in the visual system. Annu Rev Neurosci 8:547–583
Sherk H, Horton JC (1984) Receptive-field properties in the cat's area 17 in the absence of on-center geniculate input. J Neurosci 4:374–380
Shinkman PG, Bruce CJ (1977) Binocular differences in cortical receptive fields of kittens after disparate binocular experience. Science 197:285–287
Shipp S, Zeki S (1985) Segregation of pathways leading from area V2 to areas V4 and V5 of macaque monkey visual cortex. Nature 315:322–325
Shlaer R (1971) Shift in binocular diparity causes compensatory change in the cortical structure of kittens. Science 173:638–641
Skottun B, Grosof D, De Valois RL (1985) Responses of simple and complex cells in cat visual cortex to visual dot stimuli. Soc Neurosci Abstr 11:651
Spitzer H, Hochstein S (1985a) Simple- and complex-cell response dependences on stimulation parameters. J Neurophysiol 53:1244–1265

Spitzer H, Hochstein S (1985b) A complex-cell receptive-field model. J Neurophysiol 53:1266–1286

Srinivasan MV, Laughlin SB, Dubs A (1982) Predictive coding: a fresh view of inhibition in the retina. Proc R Soc Lond Biol 216:427–459

Stork DG, Levinson JZ, Albrecht DG, De Valois RL, Thorell LG (1982) Receptive fields and the optimal stimulus. Science 216:204–205 (Letters)

Tanaka K (1983) Cross-correlation analysis of geniculostriate neuronal relationships in cats. J Neurophysiol 49:1303–1318

Tanaka K (1985) Organization of geniculate inputs to visual cortical cells in the cat. Vision Res 25:357–364

Thompson P, Latchford G (1986) Colour-contingent aftereffects are really wavelength-contingent. Nature 320:525–526

Thorell LG, De Valois RL, Albrecht DG (1984) Spatial mapping of monkey V1 cells with pure color and luminance stimuli. Vision Res 24:751–769

Tolhurst DJ, Dean AF (1985) Are simple cells and complex cells distinctly different? In: Rose D, Dobson VG (eds) Models of the visual cortex. Wiley, Chichester, pp 320–325

Tolhurst DJ, Thompson ID (1981) On the variety of spatial frequency selectivities shown by neurons in area 17 of the cat. Proc Soc Lond Biol 213:183–199

Tolhurst DJ, Walker NS, Thompson ID, Dean AF (1980) Nonlinearities of temporal summation in neurones in area 17 of the cat. Exp Brain Rex 38:431–436

Torre V, Poggio TA (1986) On edge detection. IEEE Trans PAMI 8:147–163

Ullman S (1976) Filling-in the gaps: the shape of subjective contours and a model for their generation. Biol Cybern 25:1–6

van Tuijl HF (1975) A new illusion: neonlike color spreading and complementary color induction between subjective contours. Acta Psychol 39:441–445

Vautin RG, Berkley MA (1977) Responses of single cells in cat visual cortex to prolonged stimulus movement: neural correlates of visual aftereffects. J Neurophysiol 40:1051–1065

Vautin RG, Dow BM (1985) Color cell groups in foveal striate cortex of the behaving macaque. J Neurophysiol 54:273–292

von der Heydt R, Peterhans E (1986) A theory of contour representation and physiological evidence for it. Perception 15: A18

von der Heydt R, Hänny P, Adorjani C (1978) Movement aftereffects in the visual cortex. Arch Ital Biol 116:248–254

von der Heydt R, Hänny P, Dürsteler MR (1980) The role of orientation disparity in stereoscopic perception and the development of binocular correspondence. Adv Physiol Sci 16:461–470

von der Heydt R, Peterhans E, Baumgartner G (1984) Illusory contours and cortical neuron responses. Science 224:1260–1262

von der Heydt R, Peterhans E, Dürsteler MR (1986a) Spatial periodicity detectors in monkey visual cortex. (To be published)

von der Heydt R, Peterhans E, Dürsteler MR (1986b) On the neural mechanism of spatial periodicity detection. (To be published)

Webster MA, De Valois RL (1985) Relationship between spatial-frequency and orientation tuning of striate-cortex cells. J Opt Soc Am A 2:1124–1132

Westheimer G (1984) Spatial vision. Annu Rev Psychol 35:201–226

Wiesel TN, Hubel DH (1966) Spatial and chromatic interactions in the lateral geniculate body of the rhesus monkey. J Neurophysiol 29:1115–1156

Wild HM, Butler SR, Carden D, Kulikowski JJ (1985) Primate cortical area V4 important for colour constancy but not wavelength discrimination. Nature 313:133–135

Yarbus AL (1967) Eye movements and vision. Plenum, New York

Zeki S (1980) The representation of colours in the cerebral cortex. Nature 284: 412–418

Zeki S (1983a) Colour coding in the cerebral cortex: the reaction of cells in monkey visual cortex to wavelengths and colours. Neuroscience 9:741–765
Zeki S (1983b) Colour coding in the cerebral cortex: the responses of wavelength-selective and colour-coded cells in monkey visual cortex to changes in wavelength composition. Neuroscience 9:767–781
Zeki S (1983c) The distribution of wavelength and orientation selective cells in different areas of monkey visual cortex. Proc Soc Lond Biol 217:449–470
Zeki S (1984) Colour vision. In: Garlick DG, Korner PI (eds) Frontiers in physiological research. Aust Acad Sci, Canberra, pp 266–282
Zeki SM (1977) Colour coding in the superior temporal sulcus of rhesus monkey visual cortex. Proc R Soc Lond Biol 197:195–223
Zeki SM (1978a) The cortical projections of foveal striate cortex in the rhesus monkey. J Physiol (Lond) 277:227–244
Zeki SM (1978b) Uniformity and diversity of structure and function in rhesus monkey prestriate visual cortex. J Physiol (Lond) 277:273–290

Rev. Physiol. Biochem. Pharmacol., Vol. 108

Biological Activities of Complement-Derived Peptides

BERND DAMERAU

Contents

Max Planck Institute for Experimental Medicine, Department of Biochemical Pharmacology, Hermann-Rein-Straße 3, D-3400 Göttingen, Federal Republic of Germany

1 Introduction

This review deals with the biologic actions of the complement peptides C3a, C4a, C5a, and their ω-desArg derivatives. Other complement-derived split products such as C2b or Ba will not be discussed.

Historically, at the beginning of this century the first complement peptide was detected in immune complex-treated serum from guinea pigs. It was called "anaphylatoxin" since it produced anaphylactic shock-like symptoms after systemic application in the same animal species and was therefore regarded as the mediator of anaphylaxis (Friedberger 1910). This classical anaphylatoxin has recently been identified as C5a-desArg in the porcine system (Zimmermann et al. 1980; Gerard and Hugli 1981). Generation of further peptides from the third, fourth, and fifth components of complement (C3, C4, C5) was elucidated about 50 years later (Osler et al. 1959; Jensen 1967; Cochrane and Müller-Eberhard 1968; Lepow et al. 1969). Since then a wide array of biological actions has been characterized forming a mosaic-like pattern of growing knowledge, which still has to be completed in many aspects.

Complement peptides are generated in body fluids during cascading activations via two trigger pathways, the classical or the laternative, of the complement system which is a humoral defense system consisting of about 20 different proteins. In addition, they can be cleaved off selectively from their parent components by certain proteases (reviewed by Vogt 1974, 1986; Hugli and Müller-Eberhard 1978). Assuming complete conversion of the precursors maximally attainable concentrations in blood fluids of different mammalian species are $5-8 \times 10^{-6}$ for C3 peptides, about 2×10^{-6} M for C4 peptides and $3-5 \times 10^{-7}$ M for C5 peptides. By comparison, only about 20% can be reached in human lymph fluid (Vogt et al. 1986).

In general, the peptides are formed locally after unspecific or specific immunologic tissue damage and act in their proximity. Therefore, they can be regarded as autacoids or local hormones. Functionally, they are inflammatory mediators, since they activate white blood cells, smooth muscle cells, basophils/mast cells, platelets, and endothelial cells. They are involved particularly in acute inflammatory processes by their effects on unspecific defense cells, in contrast to biogenic amines (histamine, serotonin) and prostanoids which act mainly on the vascular system. According to present knowledge, the peptides initiate activation of cells by binding to specific cellular receptors. However, receptors have not yet been shown on all target cells, such as on endothelial cells, and on the target cells for their spasmogenic action in different smooth muscle organs. Recently the C5a receptor with an apparent mol. wt. of about 52000 daltons has been isolated from membranes of human granulocytes (Johnson and Chenoweth 1985).

The peptides are cationic cleavage fragments with a mol. wt. of about 9000, they consist of a single peptide chain and are stabilized intramolecularly by three disulfide bridges. Human C5a contains an additional carbohydrate moiety with a mol. wt. of approximately 3000 (for reviews see Hugli and Müller-Eberhard 1978; Hugli 1984). The primary peptides C3a, C4a, and C5a end with a C terminal arginyl residue which is cleaved off in vivo within seconds by ubiquitious monocarboxypeptidases, for instance of the N type in blood or of the B type in pancreas (Müller-Eberhard et al. 1971; Vallota and Müller-Eberhard 1973). The secondary ω-desArg derivatives, such as C3a-desArg, are reduced in or even devoid of biologic activities in most instances, but have much longer half-lives (several minutes) in vivo. They are eliminated from the circulation in various highly vascularized organs. C5 peptides are removed predominantly by specific cellular mechanisms – presumably by binding to receptors and by subsequent receptor-mediated endocytosis – in polymorphonuclear (PMN)-leukocyte-dependent (spleen) and PMN-independent (lung) ways (Glovsky et al. 1978; Webster et al. 1982).

Although chromatographic procedures to highly purify these agents have been developed (see for instance Vogt 1968; Vallota and Müller-Eberhard 1973; Hugli et al. 1981a; Manderino et al. 1982), unfractionated activated blood fluids containing complement peptides or partially purified peptide preparations have been used in numerous studies. I shall cite these studies only if they present important findings.

2 Effects on Isolated Cells and Organs

2.1 Polymorphonuclear Leukocytes

Complement peptides exert their main proinflammatory effects on PMN, which are the most important cells in acute inflammatory reactions. PMN usually play a beneficial role, but may also become deleterious when they participate in destructive diseases such as rheumatoid arthritis. Furthermore, the potential role and the pathophysiological consequences of complement peptide/PMN interactions largely depend on their localization: upon intravascular formation, the peptides cause detrimental effects by inducing bulky margination and formation of PMN aggregates. These may occlude small blood vessels and subsequently damage endothelial cells. The emerging circulatory disturbances have been proposed as contributing to the generation of adult respiratory distress syndrome (ARDS), myocardial infarcts and Purtscher's disease (for reviews see Jacob et al. 1980; Vogt 1986). On the other hand, after extravascular formation PMN are activated to exert phlogistic functions. In the next chapters, the effects of complement peptides on PMN are described in the chronological order of the cell's life cycle, i.e., starting with possible influences on PMN mobilization from bone marrow, passing to stimulation in the blood stream (aggregation, margination/adhesion), and ending with the effector functions exocytosis, oxygen radical formation, and phagocytosis.

2.1.1 Leukopoiesis and Mobilization from Bone Marrow

Whether complement peptides affect leukopoiesis has not yet been investigated, except for a study of Czarnetzki and Pawelzik (1983). There, porcine C3 and C5 peptides had no effect on proliferation of eosinophils in guinea pig bone marrow cultures. On the other hand, they seem to stimulate mobilization of mature PMN from marrow in vivo: intravascular complement activation (McCall et al. 1974) or systemic application of purified peptides (Damerau et al. 1978b; Bass et al. 1980; Webster et al. 1982) induce long-lasting rises in numbers of circulating PMN after initial neutropenia. Since the changes are much higher than attainable solely by detachment of marginating cells from vessel walls, mobilization from the bone marrow must be involved.

2.1.2 Autoaggregation

Intravascularly, C3 and C5 peptides evoke formation of PMN aggregates which plug pulmonary blood vessels and may cause cardiopulmonary dysfunction (Craddock et al. 1977a, b; Hammerschmidt et al. 1978a, b, 1981). The potential pathophysiologic role of PMN aggregation in disease states will be discussed in Sect. 4.2, the aggregation process itself has been recently reviewed in more detail by Damerau and Vogt (1982).

In vitro, aggregation is measured with isolated PMN by nephelometric methods and the Coulter counter technique. Most leukoattractants (such as C3a, C5a, and synthetic *N*-formylated oligopeptides) have been found to produce this response, provided that the cells are suspended in sufficient concentrations of at least 5×10^6 particles per milliliter in medium containing Ca^{2+}, Mg^{2+} and heated close to normal body temperature; and provided that they are stirred continuously for cell contact (Craddock et al. 1977c; O'Flaherty et al. 1977b, c, 1979b; Damerau et al. 1980a). Chelation of the divalent cations with ethylenediaminotetraacetate (EDTA) prevents aggregation and leads to decomposition of already formed aggregates (Camussi et al. 1980; own unpublished observations). Aggregation is accompanied by increases of cell volume (O'Flaherty et al. 1977c, 1978b). Aggregates are rapidly formed after application of the stimulus, reach maximal number and size (20–30 PMN in vitro) after 2–4 min and may partly deaggregate thereafter (O'Flaherty et al. 1977c, 1978b; Craddock et al. 1979; Damerau et al. 1980a). Beside PMN, monocytes but not lymphocytes and myeloblasts respond to complement peptides and other chemoattractants (Craddock et al. 1977c; O'Flaherty et al. 1978b).

C3a and C5a elicit aggregation via specific receptors (O'Flaherty et al. 1979a; Damerau et al. 1980a). Both, human and porcine C5 peptides have been found to be active (Camussi et al. 1980; Damerau et al. 1980a, b). Porcine C5a is about three times as potent as its desArg derivative (thresholds of about 1×10^{-9} M and 5×10^{-9} M). For human C5 peptides quantitative data of their aggregating potency are lacking. C3a of human and porcine origin also causes aggregation but with much lower potency than the C5 peptides; porcine C3a is three orders of magnitude less active than C5a-desArg from the same species (O'Flaherty et al. 1977b; Kreutzer et al. 1978; Damerau et al. 1980a, b). Porcine C3a-desArg inconsistently causes aggregation of PMN with a threshold similar to that of C3a buth with much lower intrinsic activity (own unpublished observations).

In contrast to pronounced influences on surface adhesion and chemotaxis, blood proteins do not significantly modify the aggregating effect of porcine C5a-desArg (Damerau and Keller 1983). Therefore, changes of the plasma membrane's physicochemical properties (e.g., charge density) appear not to affect aggregability. However, cytoskeleton structures are involved in the aggregation response. Like exocytosis and formation of oxygen-containing species, aggregation is markedly augmented by cytochalasin B which induces disintegration of microfilaments, but is reduced by the antimicrotubule agent colchicine (Craddock et al. 1978; O'Flaherty et al. 1977b; Damerau et al. 1980a). Furthermore, intracellular disulfide groups and trypsin/chymotrypsin-like proteases seem to participate in the aggregation response (Damerau et al. 1980a, 1984).

Although release of arachidonic acid and its conversion to biologically active products has been earlier postulated to mediate or facilitate PMN aggregation (O'Flaherty et al. 1979c, 1981a), lipoxygenase and cyclooxygenase products seem not to be required for the action of complement peptides on human PMN. Neither the lipoxygenase inhibitors BW 755C and nordihydro-guaiaretic acid (NDGA) (results with NDGA are to be published) nor the cyclooxygenase inhibitors indomethacin and phenylbutazone reduce the effect of C3a, C5a, and C5a-desArg at drug concentrations which are specific for inhibition of the respective target enzymes (Damerau et al. 1982). In contrast to these findings, indomethacin and ETYA (5,8,11,14-eicosatetraynoic acid), a dual cyclo-oxygenase and lipoxygenase inhibitor, have been reported to impede aggregation (O'Flaherty et al. 1979d; Camussi et al. 1981a). However, since in the latter studies medium without serum albumin was used, the concentration of free, active drug should have been markedly higher than in our studies with medium containing 0.5% human serum albumin (HSA). Therefore, inhibition may have been due to unspecific drug effects.

Beside lipoxygenase products, platelet activating factor (PAF) has been suggested as the final common mediator of aggregation (Camussi et al. 1981a). However, the observed concomitance of PMN aggregation and PAF release, as well as inhibition of these responses by the same drugs does not prove a causal relationship. Furthermore, lack of cross-deactivation in exocytosis points to independent actions of both agents (Shaw et al. 1980).

In contrast to platelet aggregation, PMN aggregation seems to be relatively insensitive to many drugs. In my opinion, drugs which inhibit the Ca^{2+}-dependent phase of PMN activation, such as calcium channel blockers and calmodulin antagonists, are possible candidates for effective inhibition. Beyond this, drugs increasing intracellular levels of cyclic adenosine monophosphate (cAMP) (e.g., isoproterenol, prostaglandin $(PGE)_1$, prostacyclin $(PGI)_2$, dibutyryl cAMP) have been claimed as impairing aggregation (Camussi et al. 1981a, b). Casting doubt on these findings, we have not been able to reduce aggregation of human PMN even by relatively high concentrations of a stable PGI_2 analogue (ZK 36374, tested up to 10^{-6} M; unpublished observations).

Aggregation due to different agents is diminished by certain corticosteroids in very high concentrations (millimolar range) only. Hydrocortisone and methylprednisolone are inhibitory in vitro and in vivo, whereas dexamethasone is much less active (O'Flaherty et al. 1977a; Hammerschmidt et al. 1979).

2.1.3 Adhesion, Margination, Leukopenia

In addition to autoaggregation, intravenous application of complement peptides causes massive PMN margination (i.e., sticking to vessel walls) particularly in lung circulation, whereas after their generation throughout the entire blood stream PMN marginate all over the microcirculation (O'Flaherty et al. 1978a). During the leukopenic reaction, which is due to aggregation plus margination, numbers of circulating neutrophils, eosinophils, and monocytes, but not of lymphocytes, are markedly decreased (McCall et al. 1974; Craddock et al. 1977b, c; O'Flaherty et al. 1977c, d; Bass et al. 1980). Leukopenia and sequestration of PMN in pulmonary vessels are transient: due to detachment of adhering cells, previous cell concentrations are reached again after 10–20 min, except for prolonged falls in eosinophil numbers (Bass et al. 1980) or when the stimulus is continuously applied for longer time periods (Fehr and Jacob 1977; Gee et al. 1985). Fehr and Jacob (1977) were the first to postulate a functional relation between complement peptide-induced margination and PMN adhesion to artificial surfaces. The latter phenomenon has often been used for studies on adhesion because of its experimental handiness.

In vitro, PMN adhesion to glass is augmented by C5 peptides after short contact (Smith et al. 1979; Smith and Hollers 1980; Damerau et al. in preparation). After surface contact for more than 15 min, spontaneous adhesion may attain or even surpass the stimulated response in which cells often adhere less firmly because of enhanced migration (Damerau et al. in preparation). These time-dependent changes are most likely the reason for earlier negative findings for C5 peptides (Keller et al. 1981; Damerau and Keller 1983). In contrast to C5 peptides, porcine C3 peptides appear to inhibit the adhesion response (Damerau et al. in preparation). Furthermore, prior incubation of suspended PMN with C3a, C5a, and FMLP (N-formyl-methionyl-lencyl-phenylalanine) produces sustained increases of adhesiveness to artificial surfaces, even in the absence of the conditioning agent (Smith et al. 1979; own unpublished observations), whereas it does not promote and even abolishes autoaggregation and adhesion to endothelium due to specific deactivation. A further difference between intercellular adhesion and attachment to artificial surfaces is that adhesion to glass is impeded by serum/plasma and by serum albumin, whereas autoaggregation is not affected (Damerau and Keller 1983). In addition, adhesion to glass is not as susceptible to inhibitory actions of certain drugs as sticking to endothelium (Fricke et al. 1985b). Therefore, PMN attachment to different surfaces is regulated by different mechanisms. Fehr and Huber (1984) have recently denied a physiologic role of C5 peptides in the adhesion response, at least to artificial surfaces. They have concluded from their experiments that activated human plasma evokes adhesion and exocytosis by principles other than C5a and C5a-desArg, presumably by components of the alternative complement pathway.

To mimic margination in vivo more closely, interactions between PMN and endothelium have been extensively studied in recent years. C5 peptides, FMLP, and LTB_4 stimulate PMN adhesion to cultured endothelials cells (Hoover et al. 1980, 1984; Gimbrone et al. 1984; Tonnesen et al. 1984). However, responses have been measured under static, not floating conditions so that the spontaneous adhesion was unphysiologically high (3–5 PMN per endothelial cell; Gimbrone et al. 1984; Tonnesen et al. 1984). Furthermore, culturing in itself presents the danger of altering relevant cell properties, e.g., PGI_2 production.

To avoid such disadvantages, guinea pig (and human) PMN have been superfused over freshly prepared autologous aortic strips (Fricke et al. 1985a). In this model, procine C3a, C5a, and their desArg derivatives increase adhesion markedly, C5a being most potent with a threshold of less than 10^{-10} M (in contrast to equal potency of human C5a and C5a-desArg, found under static conditions by Tonnesen et al. 1984). The other three agents are approximately 100 times less active (Otte et al. 1985). All peptides act primarily on PMN, since they, but not the endothelium, are deactivated by pretreatment with C3a or C5a (Tonnesen et al. 1984; Fricke et al. 1985a; Otte and Damerau 1986). Surprisingly, pharmacologic analysis has revealed a complex biochemical cooperation between PMN and endothelium in the C5a-desArg-induced response (Fricke et al. 1985a): most likely, PMN upon stimulation express a lipoxygenase which is nonoperative in resting cells (in accordance with the findings of Clancy et al. 1983), and then metabolize endothelium-derived arachidonic acid into a lipoxygenase product, presumably LTB_4. This principle causes the final adhesion response. In concert with this concept, LTB_4 has been found to augment PMN/endothelial adhesion (Gimbrone et al. 1984; Hoover et al. 1984; Fricke et al. 1985a). Furthermore, the induced response is suppressed by various lipoxygenase inhibitors (Fricke et al. 1985a; Damerau et al. 1986b). Accordingly, BW 755C (but not the cyclooxygenase inhibitor ibuprofen) has recently been shown to attenuate the neutropenic reaction in rabbits induced by C5a-desArg-containing activated plasma. In addition, the inhibitory effect of BW 755C is considerably augmented by the calcium channel blockers verapamil and nifedipine, indicating that blockade of Ca^{2+} influxes further impairs the cooperative mechanism required for margination (Issekutz et al. 1985). Increased liberation of arachidonic acid from endothelium, one of the prerequisites for the proposed model, has already been demonstrated. Porcine C5a and C5a-desArg, but not C3 peptides from the same species, stimulate both release of arachidonic acid and subsequent production of PGI_2 from rabbit isolated aorta (Rampart et al. 1983). Substrate mobilization from endothelium may be further augmented upon contact with activated PMN via reactive oxygen derivatives (Rampart et al. 1985). Although porcine C3a and C3a-desArg do not enhance PGI_2 production in aortic strips, they may also utilize the lipoxygenase pathway, since their action is blocked by the lipoxygenase inhibitor gossypol (Otte, in preparation).

Endogeneous TxA_2 and E/F-type prostaglandins seem not be involved in the response to the physiologically more important C5 peptides. Inhibitors of the respective synthetases are virtually ineffective (Fricke et al. 1985a). However, PGI_2 may modulate PMN/endothelial interaction, for endogenous endothelial PGI_2 has been found to impede adhesion of unstimulated PMN (Zimmerman et al. 1985); exogeneous PGI_2/-analogue diminishes spontaneous and induced responses in vitro (Fricke et al. 1985a), as well as C5a-desArg-evoked neutropenia in vivo (Camussi et al. 1981b).

Since in the aforementioned response to C5 peptides simultaneous stimulation of and close contact between both cellular reactants are required for the biochemical cooperation, this type of margination may ensue only when the stimulus acts from the luminal side of blood vessels, i.e., after systemic infusion or intravascular generation. Therefore, it may hold for widespread margination during acute leukopenic reactions, but not for local margination in inflammatory processes where chemoattractants which have been formed at extravascular sites are most likely diluted by the blood stream too rapidly in order to activate circulating PMN. The latter type of (inflammatory) adhesion has also been demonstrated with the superfusion model of Fricke et al. (1985a): in the absence of PMN, porcine C5 peptides substantially increase the adhesiveness of guina pig aortic endothelium within a few minutes, whereas C3a and C3a-desArg are inactive in this respect. Attachment of unstimulated PMN is then augmented for at least 30 min at 37°C, even after removal of the stimulus (Otte et al. 1985). This endothelial conditioning seems to be caused by endogenous cyclooxygenase products (TxA_2?), as it is prevented by prostaglandin synthetase inhibitors (Otte, in preparation). Selective conditioning of endothelium (Hoover et al. 1984), of PMN (Tonnesen et al. 1984) or of both cell types (Hoover et al. 1980) has been reported to be produced by C5 peptides, LTB_4, and FMLP. However, since in those studies adhesion has been assessed under nonphysiologic circumstances (cultured endothelium, static conditions), some findings may be due to artifacts. For instance, priming of PMN which has been found in attachment to artificial surfaces (see above) may have occurred in cell cultures due to contact with the surface of the culture dishes.

Whether endothelial conditioning is sufficient to trigger PMN emigration in vivo – without the need for a chemoattractant's concentration gradient – cannot yet be answered. Support for such a comparatively simple mechanism comes from recent studies on the regulation of acute inflammatory processes in rabbit skin (Colditz and Movat 1984a, b). These authors conclude from their data that PMN influx into lesions is not controlled solely by concentration gradients of chemoattractants but rather by the endothelial cells.

2.1.4 Locomotion

Complement peptides, predominantly C5a and C5a-desArg, evoke both directed (chemotactic) and undirected random (chemokinetic) migration of neutrophil, eosinophil, and basophil PMN, furthermore of monocytes/macrophages and certain tumor cells (Kay et al. 1973; Lett-Brown et al. 1976; Ogawa et al. 1981; details and references for the latter two groups of cells are given in chapter 2.2). (Chemotaxis means vectorial, directed migration of cells approaching the site of a chemoattractant's generation by moving up the concentration gradient, whereas chemokinesis means enhanced random migration). Albeit chemotaxis can be measured in vitro only, it has usually been assumed to be the main mechanism for leukocyte emigration from blood vessels and accumulation at local inflammatory sites in vivo. However, convincing evidence for this relation has not yet been provided.

As a further sign for activation of the locomotory apparatus, the peptides transform the normal spherical shape of PMN into an elongated polarized configuration. This response can be engendered in adhering as well as suspended cells. It is independent of extracellular Ca^{2+} and Mg^{2+}, but is prevented by cytochalasin B (Smith et al. 1979; Showell et al. 1982b; Shields and Haston 1985).

In brief, the history of identifying the complement peptides as physiologically important chemotactic agents startet with the design of a reliable and easy method to measure chemotaxis, the Boyden chamber, and the observation that strong chemotactic activity is generated in blood fluid under conditions which lead to complement activation, such as treatment with immune complexes (Boyden 1982). The formation of such activity in plasma from several mammalian species indicated is general biologic significance (Keller and Sorkin 1965). Although the chemotactic effect was first erroneously ascribed to the C56 complex (Ward et al. 1965), it was later identified to be due to C3a and C5a anaphylatoxins (Cochrane and Müller-Eberhard 1968; Shin et al. 1968). C5 peptides, particularly C5a-desArg because of its much longer half-life in vivo, have been suggested to be the main principles in activated plasma and serum (Chenoweth and Hugli 1980). By contrast, data on their most relevant source in inflammatory lesions, namely the lymph fluid filling the extravascular space, are sparse: Vogt et al. (1986) have found that complement peptides are formed in lymph in much smaller amounts than in blood, and that production is considerably augmented by stimulated leukocytes.

Chemotactic actions on PMN have been unambiguously demonstrated for C5 peptides from various mammalian species such as man, pig, rabbit, and guinea pig (Shin et al. 1968; Snyderman et al. 1970; Fernandez et al. 1978; Damerau et al. 1980b). Primary human and porcine C5a have thresholds of about 10^{-10} M and are usually 10–50 times more potent than C5a-desArg (Fernandez et al. 1978; Chenoweth and Hugli 1980; Damerau et al. 1980b; Webster et al. 1980a). Under certain conditions in vitro, human C5a-desArg may be devoid of chemotactic activity in the absence of a serum "helper factor" (Fernandez et al. 1978).

In contrast to the well-documented activity of C5 peptides, that of C3 peptides is controversial. Activity is distinctly species-dependent: whereas human C3a has been found to be ineffective (Fernandez et al. 1978), porcine C3a and C3a-desArg are active, but are about 100 times less potent than C5a-desArg from the same species. In contrast to C3a, C3a-desArg is only inconsistently chemotactic (Damerau et al. 1978a, 1980a). In addition, human C4a (tested in concentrations up to 5×10^{-6} M) has been found to lack chemotactic activity (Gorski et al. 1979).

Further degradation by proteases leads to smaller fragments of the complement peptides and alters their biologic effectiveness. Hence, such proteases might play a regulatory role in certain pathologic processes. Destruction of their native conformations by trypsin- and chymotrypsin-like proteases has for long been known to abolish completely the intrinsic activity of C5 peptides but only to diminish that of C3a. Residual activity in the latter case is due to C terminal fragments. Furthermore, removal of a few amino acids from the C terminal region by carboxypeptidases destroys the receptor-activating domain of complement peptides, also resulting in complete loss of activity (reviewed by Vogt 1974, 1986; Hugli 1978, 1984). By contrast, shortening of the N terminal region, for instance by aminopeptidases, affects potencies differently, suggesting highly selective interactions of this with the C terminal region. As regards chemotactic activity, porcine C5a-desArg has to rely on an intact N terminal portion, much more than C3a and C5a from the same species (Damerau et al. 1986a).

In vitro, but presumably not in vivo (Colditz and Movat 1984b), C5 peptides are active in relatively narrow concentration ranges of about two orders of magnitude only. At supraoptimal stimulus intensity chemotaxis gradually declines to control levels (Fernandez et al. 1978; Chenoweth and Hugli 1980; Damerau et al. 1980b; Webster et al. 1980a). From this, bell-shaped dose-response curves result. The reason for the "high-dose inhibition" is not definitely known: directed migration may be impeded by increased cell adhesiveness as suggested by Webster et al. (1980b), by impaired direction finding and/or by cellular deactivation (Keller et al. 1977). In contrast to C5 peptides, porcine C3a and C3a-desArg do not show this characteristic (Damerau et al. 1980a). Accordingly, C3a does not render suspended rabbit PMN unresponsive to subsequent chemotactic stimulation with the same agent (own unpublished observations).

Tachyphylaxis to C5 peptides and other chemoattractants has been proposed as being a major regulatory mechanism by which "unnecessary" or continuous cell stimulation at constant stimulus intensity is prevented (Meuer et al. 1982). Loss of responsiveness appears to be predominantly due to receptor down-regulation after ligand binding, which is especially firm with C5 peptides (Chenoweth and Hugli 1978). Deactivation is therefore essentially stimulus specific but may become partially unspecific at high ligand concentrations (Donabedian and Gallin 1981). Furthermore, several external factors may modify the deactivation process, possibly by changing ligand-receptor interactions and subsequent receptor processing. For instance, PMN adhering to glass surfaces exhibit stimulated migration for long periods of time without any loss of reactivity, whereas suspended cells are rapidly deactivated by C5 peptides and FMLP (Damerau et al. 1985a). On the other hand, during inflammatory processes in vivo (sepsis, infectious, and non-infectious skin diseases) PMN dysfunctions have been ascribed to prior contact with C5 fragments (Solomkin et al. 1981), but selective unresponsiveness to C5a may also be induced or aggravated by other, not yet characterized, factors (Schröder and Christophers 1985).

Despite a considerable number of data on chemotaxis, the definitive identification of an individual complement peptide as a chemoattractant (and even more the assessment of its biologic significance for leukocyte accumulation in vivo) is hampered by the following difficulties:

1. Different chemotaxis assays may give diverging results
2. Proteins often added to suspension media may modify migration, mainly by changing cell adhesion to chemotaxis filters and other artificial surfaces used in the assays
3. Certain complement peptides may require specific cofactors for optimal activity
4. Chemotactic responses may be modified by different extents of binding of chemottractants to the surfaces on which cells move

In the following section, these potential difficulties will be discussed in the same order.

First, sensitivity of chemotaxis assays can vary markedly: Kreutzer et al. (1978) have found that in the under-agarose assay complement peptides are about 100 times less active than in the Boyden chamber assay.

Secondly, for most assays in vitro proteins, e.g., serum albumin or gelatine, are essential; they lower cell adhesion to a level at which detachment and hence migration become possible. These proteins already produce some chemokinetic migration (Keller et al. 1978; Hakansson and Venge 1985) which may be enhanced by chemoattractants. In the absence of a concentration gradient, complement peptides have been shown to act chemokinetically (Damerau et al. 1978a; Webster et al. 1980a). Whether chemokinesis plays any role in PMN locomotion in vivo, is not known. Nevertheless, it should be determined in order to quantify the chemotactic activity of an agent as exactly as possible. However, at least in Boyden chamber assays complement peptides usually cannot be distributed evenly in the filters, a difficulty that is due to strong adsorption to the filter matrix (Damerau et al. 1983). Therefore, data on chemokinesis may be erroneous, in particular those obtained with the often-used checkerboard assay.

Thirdly, beside unspecifically acting chemokinetic agents, other factors may selectively render certain complement peptides chemotactic or may facilitate their action. The in vivo significance of these principles has not been elucidated. The first-described cofactor ("cochemotaxin"), a cationic peptide purified from rat serum, was claimed to be necessary for the action of C5a-desArg from pig, guinea pig and rat (Wissler et al. 1972). In contrast to these findings, in our laboratory the chemotactic action of porcine C5a-desArg does not depend on any specific cofactor (Damerau et al. 1980b). From human plasma/serum a heat-stable, anionic polypeptide with a mol. wt. of approximately 60 000 has been isolated. This helper factor or cochemotaxin augments markedly the chemotactic effect of human C5a-desArg, but has no activity on its own nor a facilitating influence on primary C5a (Fernandez et al. 1978; Perez et al. 1980, 1981). However, in another study human C5a-desArg has been found to be active in an environment completely free of protein. This suggests that the helper factor is required under special technical conditions only (Chenoweth and Hugli 1980). As a further modifying agent, C1 inhibitor (C1 INH) has been reported to increase chemotaxis and, in addition, to prevent deactivation of PMN by human C5 fragments (Patrick et al. 1980). On the other hand, natural inhibitors of the chemotactic response have also been detected: a cationic protein, which is present in serum of patients with systemic lupus erythematodes and is immunologically related to the Bb fragment of complement factor B

(Perez and Hooper 1986), inhibits selectively chemotactic – but not exocytotic or aggregating – effects of human C5a-desArg. This factor does not affect actions of other chemoattractants (Perez et al. 1978, 1980).

Fourthly, quantitating interpretations of data from chemotaxis assays may be complicated by strong adsorption of complement peptides to the matrix of chemotaxis filters (Webster et al. 1980b). Furthermore, (porcine) C3 and C5 peptides are bound to markedly different degrees (Damerau et al. 1983). As the adsorbed peptide material produces a normal chemotactic response, statements on active concentrations and shapes of concentration gradients are difficult. They are certainly wrong if based on the prevailing diffusion models. Regarding the possible biologic role, Webster et al. (1980b) have proposed that PMN migration in vivo may be directed to a considerable extent by surface-bound chemoattractants. This should form concentration gradients which are more stable than those due to diffusion. However, evidence for this mechanism in vivo has not yet been presented.

As regards biochemical and pharmacologic analysis of chemotaxis, the great majority of studies have been performed with stimuli other than complement peptides, e.g., with synthetic N-formylated oligopeptides. The data obtained are too numerous to be summarized here. As far as parameters of synthetic and complement peptides have been directly compared, similar mechanisms have been found in signal transfer and cell activation (reviewed for instance by Snyderman and Pike 1984). Binding of ligands to specific receptors triggers changes in transmembrane potential, ion fluxes, levels of cyclic nucleotides, cytoskeleton structures, and in membrane lipids with release of arachidonic acid and subsequent formation of cyclooxygenase/lipoxygenase products, as well as activation of protein kinase C, transmethylation reactions, and increased phosphatidylinositol turnover. A model for the biochemical events has recently been proposed by Snyderman and Pike (1984).

Specifically, the chemotactic action of complement peptides has been shown to be optimal in the presence of extracellular Ca^{2+}, Mg^{2+}, K^+, and Na^+ (Becker and Showell 1972; Showell and Becker 1976). Since the facilitating effect of external K^+ is blocked by ouabain, an Na^+/K^+-dependent ATPase may be involved (Showell and Becker 1976). Intracellular cAMP is transiently elevated after stimulation with chemoattractants but does not quantitatively correlate with the chemotactic response (Naef et al. 1982). Therefore, it may not be a second messenger in chemotaxis. On the other hand, drug-induced sustained rises of intracellular cAMP diminish PMN responsiveness (Rivkin et al. 1975).

Endogenous arachidonic acid metabolites seem to be negligible for the chemotactic response. Although Goetzl (1980) and Showell et al.

(1980) have reported that lipoxygenase inhibitors attenuate C5 peptide-induced chemotaxis, the concomitant decrease of spontaneous random migration in the absence of a stimulus points to unspecific mechanisms. In concert with this conclusion, we have not found BW 755C and NDGA to be inhibitory. These agents even augment C3a/C5a-evoked directed migration (Damerau et al. 1982). As indomethacin does not affect the response, cyclooxygenase products, also, seem not to be secondary mediators (Goetzl 1980; Damerau et al. 1982).

2.1.5 Exocytosis

Complement peptides induce release of lysosomal enzymes (exocytosis) with preferential liberation of storage products from the secondary specific granules, e.g., of lysozyme, lactoferrin, and vitamin B_{12} binding protein. Release from these granules ensues at a lower level of stimulation and reaches higher amounts than that from the primary lysosomes (Henson et al. 1978; Wright and Gallin 1979; Showell et al. 1982a; Naef et al. 1984). However, optimal exocytosis is achieved only in the presence of cytochalasin B (Goldstein et al. 1973; Henson et al. 1978; Webster et al. 1980a; Showell et al. 1982a; Naef et al. 1984), or when either the stimulating agent or the cells adhere to a surface, for instance during migration into chemotaxis filters (Becker et al. 1974; Wright and Gallin 1979). Without these prerequisites, lysosomal release is limited or even undetectable (Henson et al. 1978; Showell et al. 1982a; Naef et al. 1984).

Special case of exocytotic action may be C3a-induced secretion of mucous glycoprotein from human isolated airways. C3a-desArg is inactive in this respect. The secretory response seems to be independent from histamine and lipoxygenase products (Marom et al. 1985). The authors speculate on its probable role for bronchial asthma. However, it is quite unlikely that intact C3a can reach sufficient concentrations at this site, without being inactivated by carboxypeptidases.

Human C5 peptides are active on human and guinea pig neutrophils at concentrations similar to those in other biologic responses. Threshold of C5a is about 10^{-10} M, of C5a-desArg about 10^{-8} M (Henson et al. 1978; Chenoweth and Hugli 1980; Webster et al. 1980a; Gerard et al. 1981), whereas the thresholds of porcine C5 peptides are about one order auf magnitude higher (Gerard et al. 1981). Human C5a seems to be ineffective on guinea pig eosinophils (Ogawa et al. 1981). Human C3a is about 100 times less active on human neutrophils than homologous C5a-desArg (Showell et al. 1982a). Casting doubt on the exocytotic effect of C5 peptides, Fehr and Huber (1984) have recently found only marginal release from suspended human PMN by C5a-desArg-containing activated human plasma, even in the presence of cytochalasin B. By contrast, adhering PMN liberate marked amounts of secondary granule contents upon stimulation. However, the response is apparently due to a diisopropyl fluorophosphate (DFP)-inhibitable, heat-labile principle, rather than to C5a-desArg. This principle seems to originate from the alternative complement pathway and is presumably identical with factor Bb. Nevertheless, these results do not necessarily hold for extravascular sites where complement peptides usually operate.

The exocytotic response (reviewed by Becker and Henson 1973; Weissmann et al. 1980, 1981) is specific, energy requiring and noncytolytic, not causing loss of lactate dehydrogenase (Goldstein et al. 1973; Becker et al. 1974; Henson et al. 1978; Showell et al. 1982a). It utilizes extracellular as well as intracellular Ca^{2+} ions (Shaw et al. 1982), whereas Mg^{2+} is without influence (Goldstein et al. 1975a). For an optimal response, external K^{+} and Na^{+} are obligatory (Showell et al. 1977). In contrast to a slow release by A 23187 and a phorbol ester, the response to complement peptides reaches its maximum after 1–2 min at 37°C in the presence of cytochalasin B (Goldstein et al. 1973; Henson et al. 1978; Webster et al. 1980a). It is considerably retarded if PMN adhere to chemotaxis filters (Becker et al. 1974). Unlike the slow, prolonged enzyme release from rabbit alveolar macrophages by C5 peptides, it does not require protein synthesis (McCarthy and Henson 1979; Henson et al. 1981b). PMN are rapidly deactivated when the stimulus is applied sufficiently long before cytochalasin B or Ca^{2+} and Mg^{2+}. Deactivation is stimulus specific, indicating that only early steps – receptor occupation and possibly signal transfer – are affected (Henson et al. 1978), whereas partial cross-deactivation between C5a, FMLP, and a chemotactic bacterial lipid has also been found (Henson et al. 1981a).

Endogenous lipoxygenase products may be involved in exocytosis, particularly in the response augmented by cytochalasin B (Showell et al. 1982a), since it is reduced by lipoxygenase inhibitors such as ETYA, NDGA and quercetin (O'Flaherty et al. 1979d; Showell et al. 1980). Nevertheless, neither LTB_4 nor PAF seem to be major mediators, as neither agent which induces selective deactivation renders human PMN unresponsive to C5a and other stimuli (O'Flaherty et al. 1981b, c).

However, sustained elevations of intracellular cAMP may impede the exocytotic response (Goldstein et al. 1973). Currently, the idea of PMN activation by protein kinase C-dependent mechanisms is favored. In concert with such models (proposed for instance by Snyderman and Pike 1984) are findings that diacylglycerols, which can directly activate this enzyme, enhance C5a-evoked exocytosis (O'Flaherty et al. 1984), but data raising doubts about an essential involvement of protein kinase C in induction of exocytosis and respiratory burst by complement peptides and *N*-formylated peptides have been recently published (Gerard et al. 1986; Wright and Hoffman 1986). Possibly at a previous step, release and chemotaxis are blocked by pertussis toxin which seems to interact with a guanine nucleotide regulatory protein not being related to adenylate cyclase (Verghese et al. 1985).

Regarding its (patho)-physiologic role, lysosomal enzyme release from stimulated PMN has been proposed to be involved – e.g., by the action of lysosomal constituents (proteases, lipases, etc.) – in PMN chemotaxis (Gallin et al. 1978), in breaking up the migration path through basement membranes of vessel walls (Henson et al. 1981b), in microbial killing, and in digestion of foreign as well as host material. Tissue damage may essentially contribute to chronic inflammatory and degenerative processes, such as rheumatoid arthritis (Weissmann 1982), pulmonary emphysema, and ARDS (Baggiolini and Dewald 1985). Furthermore, lysosomal proteases may also participate in regulation of inflammatory processes by causing sequential generation of C5a by a constituent of specific granules and later its inactivation by a product from azurophil granules, presumably elastase (Wright and Gallin 1977)

To elucidate the significance of complement peptide-evoked exocytosis in vivo, the intensity and the functional consequences of PMN contact to natural reactants have to be studied, since surface adhesion is the only known physiologic condition rendering this response possible. In this respect, Henson et al. (1981b) did not detect secretion from stimulated PMN in contact with cultured fibroblasts, endothelial, and smooth muscle cells, in contrast to marked exocytosis from PMN adhering to artificial surfaces. Therefore, appropriate conditions for release in vivo have still to be defined.

2.1.6 Formation of Reactive Oxygen Derivatives and Phagocytosis

Treatment of PMN with a variety of stimulating agents, such as chemoattractants, calcium ionophore A 23187, phorbol esters, concanavalin A and phagocytosable particles, causes a pronounced respiratory burst with highly increased oxygen consumption, activation of the hexose monophosphate shunt and production of several oxygen-containing species. Among the latter compounds cytotoxic radicals, e.g., O_2^-, $\cdot OH$, are formed via different oxygen-metabolizing enzyme systems (reviewed by Fantone and Ward 1982; Babior 1984; Flohé et al. 1985). This formation does not depend on exocytosis, nor on phagocytosis (Goldstein et al. 1975b).

The enzyme being stimulated and triggering the chain of enzymatic reactions is a membrane-associated nicotinamide adenine dinucleotide phosphate (NADPH)-dependent oxidase which metabolizes dioxygen into superoxide anion (O_2^-). This oxidase is apparently activated via various transduction mechanisms, utilized by chemoattractants, A 23187, or phorbol esters (McPhail and Snyderman 1983; Gerard et al. 1986).

The following complement peptides have also been found to elicit the respiratory burst: human C5a has a threshold of 10^{-9} M for O_2^- generation; maximum is still not reached at more than 10^{-7} M (Dahinden et al. 1983; McPhail and Snyderman 1983; Gerard et al. 1986). C5a-desArg from the same species is approximately 50 times less potent (Webster et al. 1980a). In addition, the activity of human C5a is greatly enhanced in the presence of a yet unidentified serum factor (protein?) (Maly et al. 1983). Its possible relation to other helper factors which have been described for chemotaxis (see chapter 2.1.4) is not known.

Like exocytosis, formation of oxygen species (O_2^-, H_2O_2) becomes optimal either in the presence of cytochalasin B or when the cells adhere firmly to surfaces (Goldstein et al. 1975b; Becker et al. 1979; Dahinden et al. 1983), whereas the effects of A 23187 and a phorbol ester do not require cytochalasin B (McPhail and Snyderman 1983). In suspended PMN, C5 peptides and FMLP cause a rapid transient stimulation of the oxidase which levels off after a few minutes (Webster et al. 1980a; McPhail and Snyderman 1983). The cause of the rapid fall of O_2^- production is unknown; it seems not to be due to internalization of the oxidase together with chemoattractant receptors. Events in the transduction pathway at the level of second messengers or in the membrane environment are more likely (Parkos et al. 1985). In contrast, adhering PMN respond with a sustained O_2^- production for at least 60 min, correlating closely with the extent of adhesion (Dahinden et al. 1983). On the other hand, A 23187 and phorbol ester elicit similarly long-lasting production even in suspended PMN (McPhail and Snyderman 1983).

As in exocytosis, formation of oxygen species depends on intracellular Ca^{2+} ions, whereas extracellular Ca^{2+} enhances the response to an optimal level (Becker et al. 1979). Agents acting via specific receptors cause selective loss of responsiveness (Simchowitz et al. 1980). In addition to receptor occupation, subsequent temperature-sensitive, converging steps may be involved in deactivation as cross-deactivation between a C5 peptide and FMLP has been demonstrated (English et al. 1981). The authors propose that this nonselective deactivation may hold the potential for O_2^- production until PMN have arrived at central parts of inflamed areas. Though deactivated to soluble, chemotactic stimuli, the cells retain normal or even acquire enhanced responsiveness to phorbol esters and phagocytosable particles (English et al. 1981; Bender et al. 1983). Accordingly, the potential for production of oxygen species and for bactericidal activity can be markedly enhanced after prolonged prior contact with chemoattractants (Van Epps and Garcia 1980). The priming effect then lasts for several hours. This may be analogous to conditioning in PMN adhesiveness to artificial surfaces (see Sect. 2.1.3).

Regarding the biologic signifiance, PMN-derived oxygen species appear to be essentially involved in microbial killing and in damage to host cells (Fantone and Ward 1982). Examples for detrimental effects are endothelial lesions, which aggravate certain shock states and lung inflammation (see Sects. 4.1, 4.2), and aggregation of IgG molecules, which has been assumed to support the self-perpetuating destructive process in rheumatoid arthritis (Lunec et al. 1985). In addition, oxygen species may amplify generation of C5-derived chemoattratants (Shingu and Nobunaga 1984) and can transform foreign compounds, such as drugs, into free radical intermediates which possibly contribute to local drug toxicity and to chemical carcinogenesis (Kalyanaraman and Sohnle 1985). Such reactions should be favored in inflammatory lesions because of the huge numbers of activated PMN.

Data about the influence of complement peptides on phagocytosis are rare. It seems to be differently affected by C5a: when the peptide is adsorbed to latex particles, it promotes their engulfment (Henson et al. 1981b). In soluble form, C5a retards phagocytosis of sensitized sheep erythrocytes (Musson and Becker 1976).

2.2 Monocytes/Macrophages and Tumor Cells

Only a limited number of studies about the effects of complement peptides on monocytes and macrophages are available. The reason may lie in the traditional concept that both act primarily on different stages of defense reactions: the peptides mediating acute inflammation, the cells participating in longer-lasting or chronic inflammatory and specific immune processes. However, views have changed remarkably. Complement peptides have recently been shown to modify primary immune responses partly via macrophages (see Sect. 2.3). Monocytes/macrophages, on the other hand, secrete complement components (reviewed by Hartung and Hadding 1983), generate and inactivate complement peptides by their proteases (Snyderman et al. 1972; Kreuzpaintner et al. 1986), and are involved in regulation of acute inflammation.

In vitro, C5 peptides elicit polarization and directed migration of human peripheral blood monocytes. Both peptides from human origin are equieffective, although C5a binds more avidly to monocyte receptors than C5a-desArg. This finding is unusual, since a ligang with higher affinity is normally more potent, such as C5a in PMN chemotaxis and other actions. C5a and C5a-desArg have threshold of about 10^{-10} M and are maximally active at 4×10^{-9} M. Furthermore, a helper factor seems not to be required for the action of C5a-desArg (Marder et al. 1985). Directed migration of human monocytes is restricted to certain subpopulations. These respond largely to other chemoattractants (e.g., FMLP), too, and are specifically deactivated after prior contact with a stimulus (Falk and Leonard 1980). However, guinea pig alveolar macrophages show partial cross-deactivation between C5 fragments and FMLP in chemotaxis and O_2^- generation (Daughaday et al. 1985).

In contrast to PMN which do not re-enter the blood stream after emigration, monocytes are much more mobile (and have a markedly longer life span). Therefore, they have to be at least transiently immobilized to fulfill local functions. In vitro, the spreading reaction to artificial surfaces is viewed as a model of activation and immobilization. In this response, C5 peptides may be involved indirectly. They are presumably cleaved from membrane-associated C5 of human monocytes by factor Bb (Sundsmo and Götze 1981; Baumgarten et al. 1985). A similar mechanism has been discussed for factor Bb-induced secretion of lysosomal hydrolases from mouse peritoneal macrophages (Hirani et al. 1985). However, the contradiction that in one case C5 peptides stimulate and in the other case impede migration, cannot be resolved yet. Possibly, a concentration gradient of a C5 peptide leads to successive spreading, detachment and locomotion.

Human C5a and C5a-desArg, being equipotent, also produce secretory responses in macrophages. In contrast to exocytosis from PMN, release of β-D-galactosidase and *N*-acetyl-β-D-glucosaminidase from human and rabbit alveolar macrophages is not facilitated by cytochalasin B, but is blocked by cycloheximide. Release is very slow: it is enhanced three- to fivefold after after 72 h of incubation. In addition, C5 peptides stimulate pinocytosis of horseradish peroxidase and evoke release of proteolytic enzymes. Interestingly, production and release of a principle causing PMN chemotaxis is induced by C5a-desArg only, and not by C5a (McCarthy and Henson 1979). This may be the reason, why C5a has been found to be considerably less proinflammatory than C5a-desArg in rabbit lungs in vivo.

Procine C3a does not stimulate secretion of *N*-acetyl-β-D-glucosaminidase from elicited guinea pig peritoneal macrophages, but is itself rapidly converted to C3a-desArg by a cell-derived monocarboxypeptidase (Kreuzpaintner et al. 1986). On the other hand, secretory responses may be affected by C3 peptides, too. Human C3a-desArg has been shown to augment secretion of interleukin 1 (Laude et al. 1985), whereas porcine C3 and C5 peptides inhibit C2 production by cultured human monocytes (Lappin et al. 1983). The functional significance of these findings is not yet known. They are consistent with the idea that locally produced complement components/peptides may have regulatory functions in certain microenvironments.

Production of oxygen species in murine peritoneal and guinea pig alveolar macrophages is stimulated by relatively high concentrations of C5 peptides (Kunkel et al. 1982; Daughaday et al. 1985). Furthermore, both C3a and C5a can release cyclooxygenase products such as TxA_2, PGE_1, and PGE_2 from murine and guinea pig peritoneal macrophages (Kunkel et al. 1982; Hartung et al. 1983).

Primed monocytes/macrophages can be cytotoxic to other cells by mechanisms which are not yet completely elucidated. One of the proposed effector molecules has been C3a. It was claimed to kill various human and murine tumor cells, guinea pig and murine macrophages, as well as human lymphocytes (Ferluga et al. 1976, 1978). However, other authors did not confirm this effect (Goodman et al. 1980; Kreuzpaintner et al. 1986).

Certain C5a fragments may affect tumor cells, though in quite another way than suggested above for C3a. Treatment of human C5 with trypsin or postphagocytic PMN supernatants produces first a principle (= C5a) which attracts PMN but not tumor cells; after its destruction a smaller fragment is generated having a mol. wt. of about 6000, which is chemotactic for tumor cells only (Orr et al. 1978, 1979b). Incubation of C5a or C5a-desArg with the aforementioned protease preparations or with tumor cell extracts generates directly the latter fragment (Romualdez and Ward 1975; Romualdez et al. 1976; Orr et al. 1979a). The authors have therefore proposed that this fragment may provoke immigration of circulating tumor cells into inflamed areas and hence induce metastatic settlement. Supporting this concept, endogenous C5-derived chemotactic activity for tumor cells has been detected in exudate fluids from glycogen-induced peritonitis in rats (Orr et al. 1982). Conversely, intraperitoneal injection of this principle causes severe diffuse mesenteric metastasis of systemically applied Walker carcinosarcoma cells in the same species (Lam et al. 1981).

2.3 Immune-Competent Cells

Both C3 and C5 peptides affect the primary immune response in vitro, but it is not known whether they have similar effects in vivo. As mixed cell populations consisting of macrophages and different lymphocyte types were used in these studies, the detailed mechanisms are not yet completely elucidated. Mutual cooperation between different cell types further impedes analysis.

In brief, C3a inhibits, whereas C5a and C5a-desArg enhance the primary immune response (Goodman et al. 1982a; Morgan et al. 1982). C3a suppresses production of Fc fragment-induced, nonspecific as well as sheep erythrocyte-induced, specific antibodies. This effect can be elicited for up to 48 h after antigen application (Morgan et al. 1983a). It depends on a serum-free environment or on inhibition of serum carboxypeptidase, since the otherwise rapidly formed C3a-desArg is inactive in this respect. In the murine system, the specific antibody response is about 100 times less susceptible to the action of C3a than in the human system, with 50% inhibition at 10^{-6} M vs. 10^{-8} M C3a (Morgan et al. 1983a). The reason for the divergent sensitivities is not known. As regards the type of target cells, C3a activates nonspecific $OKT3^{+} 8^{+}$ suppressor T lymphocytes in human peripheral blood lymphocyte cultures. Furthermore, the response requires interaction of those with adherent cells (presumably monocytes). Maximal suppression is achieved already after 1–2 h of treatment with C3a (Morgan et al. 1985a). In the murine system, the effect of C3a is mediated through generation of nonspecific Lyt-2^{+} suppressor T cells and requires cooperation with an adherent cell type, possibly macrophages. In contrast, the C3a-induced suppressor T cells seem not to operate by release of inhibitory lymphokines, nor do they inhibit helper cell activity or synthesis of interleukin 2. On the other hand, interleukin 2-containing lymphokine preparations reverse the effect of C3a (Morgan et al. 1985b). Like other biologic actions, immunosuppression is evoked by the C terminal part of C3a, since synthetic C3a(70–77) is markedly inhibitory (Morgan et al. 1983b). As regards the biologic significance of these findings, Heideman and Hugli (1984) have recently proposed that enhanced generation of C3a may contribute to lowered resistance to infections in multisystem organ failure after extensive tissue injury.

On the other hand, C3a does not prevent production of antibodies to thymus-independent antigens. This is consistent with the proposed T lymphocyte-dependent action of C3a (Hugli and Morgan 1984). Furthermore, C3a does not impede tetanus toxoid- or mitogen-induced proliferation of T cells in the human system as well as Fc fragment-evoked B-cell proliferation in murine spleen cell cultures (Morgan et al. 1982). Beside its influence on the primary immune response, C3a and its C terminal analogue C3a(70–77) have been found to impair lymphokine generation ("leukocyte inhibitory factor" which inhibits PMN migration) in cultures of mitogen- or antigen-stimulated human peripheral blood monocytes plus T cells. Both peptides act in relatively low concentrations with IC_{50} of $10^{-8}-10^{-7}$ M. In addition, C3a(70–77) diminishes mitogen-induced migration of T lymphocytes (Payan et al. 1982). C3a and C3a-desArg have been reported to inhibit in vitro cytotoxicity of natural killer cells, which are spontaneously cytotoxic lymphocytes with broad target specificity against tumor cells and virus-infected cell lines (Charriaut et al. 1982).

C5a and C5a-desArg in physiologic concentrations (thresholds less than 10^{-9} M) enhance Fc fragment-mediated, nonspecific and sheep erythrocyte-induced, specific antibody production in human and murine lymphocyte cultures (Goodman et al. 1982a; Morgan et al. 1983a). Both peptides apparently trigger early steps of the immune response. Primarily, C5a receptor-bearing macrophages seem to be stimulated to secret interleukin 1, whereas in the subsequent response $1a^-$ macrophages, $1a^+$ antigen-presenting accessory cells and $OKT3^+4^+$ helper T cells have to cooperate (Goodman et al. 1982a, b). In contrast to this model, Hugli and Morgan (1984) have proposed an alternative activation mechanism which does not involve macrophages. Furthermore, C5a and C5a-desArg enhance antigen-induced T cell proliferation, but fail to influence mitogen-induced B and T cell proliferation (Morgan et al. 1983a) and, like C3a, do not affect antibody production by thymus-independent antigens (Hugli and Morgan 1984). As regards lymphocyte migration, neither C3 nor C5 peptides seem to affect it considerably at physiologic concentrations (Wilkinson 1985; Pohajdak et al. 1986).

2.4 Platelets

Sudden falls in platelet numbers in guinea pigs have previously been observed after systemic infusions of activated, peptide-containing blood fluids, but could not be definitely attributed to the action of the complement peptides (earlier literature reviewed by Giertz and Hahn 1966). First evidence for their involvement was provided by studies of Schumacher et al. (1974) and Benner et al. (1975): in cats, dextran-activated serum and partially purified C5a-desArg from rats caused transient thrombocytopenia and sticking of platelet aggregates in lung vessels. In vitro tests revealed that only feline and guinea pig platelets are aggregated, whereas cells from man, dog, rabbit, and rat do not react. Grossklaus et al. (1976) then definitely demonstrated activity for purified procine C5a-desArg with the same species dependency; in addition, platelets from pig and cattle were also resistant. They, furthermore, showed that porcine C3a causes aggregation of platelets from guinea pigs, but not from the other species. Negative results obtained with human and porcine C3a (Stimler et al. 1980) are most likely due to rapid inactivation of C3a by carboxypeptidase N present in the platelet-rich plasma used. Human C3a and C5a (or C5a in accordance with Stimler et al. 1980) elicit aggregation of washed guinea pig platelets, whereas both desArg derivatives are inactive (own unpublished observation). In contrast to earlier findings – and to present experimental evidence in our laboratory – human platelets have been claimed to be aggregated and to release serotonin upon stimulation with human C3a and even C3a-desArg (Polley and Nachman 1983).

C3a and C5a liberate ^{3}H-serotonin from guinea pig but not from human platelets (Becker et al. 1978b; Meuer et al. 1981a). When serotonin reuptake into the cells is blocked by the antidepressant drug imipramine, release becomes optimal (Meuer et al. 1981a). It is inhibited by DFP (Becker et al. 1978b). Like most other biologic responses, aggregation and release are elicited by low peptide concentrations: thresholds of C5a are less than 10^{-9} M, those of C3a (and of porcine C5a-desArg for aggregation) are 10–50 times higher (Grossklaus et al. 1976; Damerau et al. 1980b; Meuer et al. 1981a). At high concentrations, porcine C3a and C5a-desArg also liberate histamine from guinea pig platelets (own) unpublished observations). Likewise, human C4a, which acts via C3a receptors, shows weak secretory activity with a threshold of about 10^{-7} M (Meuer et al. 1981c).

C3 and C5 peptides act via specific receptors. To elicit release, C3a has been shown to bind to high- and low-affinity sites. Its effect is diminished after occupation with C3a-desArg of the low-affinity receptors (Becker et al. 1978a). By contrast, aggregation seems to be elicited via high-affinity binding only, as C3a-desArg is not inhibitory (Damerau et al. 1986). As regards structure activity relations, the receptor-activating domains apparently reside in the C terminal portions of C3a and C5a, but the N terminal parts essentially contribute to optimal potencies (Meuer et al. 1981b; Damerau et al. 1986a).

Guinea pig platelets rapidly and selectively lose their responsiveness to C3a, C5 peptides, and ADP when incubated with one of these agents under nonreactive conditions, i.e., in the absence of external Ca^{2+} or at low temperature (Grossklaus et al. 1976; Meuer et al. 1981a, b). The lack of cross-deactivation between ADP and the complement peptides indicates that endogenous ADP is not a mediator for C3a/C5a. Furthermore, in serotonin release cells are deactivated by subthreshold concentrations of C3a and C5a (Meuer et al. 1982). However, this "low-zone desensitization" seems not to be the proposed, commonly operating control mechanism preventing unnecessary cell activation as it does not occur in C3a-induced platelet aggregation and is completely absent in PMN aggregation (own unpublished observations).

Aggregation and amine release are rapid, noncytolytic, energy-dependent processes. They are highly temperature sensitive, require extracellular Ca^{2+} and glucose (Grossklaus et al. 1976; Becker et al. 1978b). Pharmacologic experiments indicate that arachidonic acid metabolites (TxA_2 ?) mediate the aggregation response (not studied for the release). However, since the action of C3a is inhibited at considerably lower concentrations of indomethacin and phenylbutazone than the response to C5a desArg, the composition and significance of the putative mediators may be different for either peptide (Damerau et al. 1982).

A biologic role of complement peptide-induced platelet activation seems to be restricted to a few, reactive animal species: only in cats and guinea pigs may it contribute to shock symptoms, after intravascular formation of complement peptides. In contrast to strong lethal effects of the C5 peptides, C3a never kills guinea pigs after intravenous administration (Grossklaus et al. 1976; Stimler et al. 1980), unless inactivation of C3a is prevented by a carboxypeptidase N inhibitor (Huey et al. 1983). In vivo, platelets from normally resistent species such as man may also be activated, namely indirectly such as by PAF derived from stimulated PMN.

2.5 Mast Cells and Basophils

Mast cells occupy a pivotal position in immediate hypersensitivity. Upon antigenic challenge they release several bioactive substances, e.g., histamine, heparin, various enzymes, PAF, and leukotrienes. Most important for acute allergic reactions is histamine. This amine contracts smooth muscles in the alimentary and respiratory tract – leading for instance to bronchospasm – affects heart rate and rhythm, lowers blood pressure by vasodilatation, and increases vascular permeability. Mast cells are activated not only by immunologic stimuli but also by many cationic substances, among them drugs and complement peptides.

Unfractionated and partially purified peptide preparations have earlier been reported to cause mast cell degranulation and histamine release (reviewed by Hugli and Müller-Eberhard 1978). Using purified peptides, Johnson et al. (1975) could demonstrate that human and porcine C3a and C5a release histamine from rat peritoneal mast cells, whereas their desArg derivatives are inactive. However, in comparison with mast cells from other species (Regal et al. 1983a), those from rat are relatively insensitive: release requires high peptide doses and reaches only 30% of total histamine content (Johnson et al. 1975). More sensitive are human basophils releasing up to 70% of their histamine (Peterson et al. 1975; Glovsky et al. 1979). Although in most experiments unfractionated peripheral blood leukocytes, consisting of a few percent basophils only, were used, the action of the peptides seems not to be affected by other cell types (Siraganian and Hook 1976). Human C5a has been found to have a threshold of about 5×10^{-9} M and to be approximately 50 times as potent as human C3a (Glovsky et al. 1979; Hartman and Glovsky 1981). In comparison with human C5a, porcine C5a-desArg is less active by about one order of magnitude (Petersson et al. 1975). Activity cannot be ascribed to human C5a-desArg, with certainty, since only unfractionated activated human serum has been tested and found to release histamine (Hook et al. 1975; Grant et al. 1976).

On both human basophils and rat mast cells combinations of C3a and C5a act synergistically (Johnson et al. 1975; Hartman and Glovsky 1981). C terminal oligopeptides of human C3a interact specifically with C3a receptors which may be a protease rapidly inactivating C3a (Gervasoni et al. 1986; Hugli 1986): the oligopeptides degranulate rat mast cells with concomitant release of histamine and serotonin and cause selective deactivation. However, they are much less potent than native C3a. Because of their small molecular size, the response to C3a and its analogues seems not to depend on receptor cross-linking such as in IgE-elicited release (Hugli and Erickson 1977; Glovsky et al. 1979). Of course, C3a and the active C5 peptides specifically deactivate mast cells and basophils. Cross-deactivation with IgE-mediated release does not occur (Petersson et al. 1975; Siraganian and Hook 1976).

Peptide-induced histamine release is an energy-requiring and noncytotoxic secretory process (Grant et al. 1976; Hook and Siraganian 1977) and is highly dependent on temperature and extracellular Ca^{2+} (Johnson et al. 1975; Petersson et al. 1975; Siraganian and Hook 1976). In contrast to the slow release from sensitized cells after antigenic challenge, it proceeds rapidly and is complete after 1–2 min (Petersson et al. 1975; Siraganian and Hook 1976; Glovsky et al. 1979). Data on its temperature dependence are contradictory: whereas release by C5a-desArg-containing, zymosan-activated serum has been found to be maximal between 25–30°C (Siraganian and Hook 1976), purified C3a has shown optimal activity at 37°C (Glovsky et al. 1979). Siraganian and Hook (1976) proposed a concomitant inactivation process, biochemical or due to receptor occupation, which works more effectively at higher temperatures and thus counteracts the release process.

Biochemical and pharmacologic analyses have revealed that complement peptide-induced responses are similar in several aspects to histamine liberation caused by IgE, phorbol esters and calcium ionophore A 23187; e.g., the release is enhanced by phosphatidylserine (Johnson et al. 1975). It is, on the other hand, inhibited by antimicrotubule agents and by drugs which augment the level of intracellular cAMP (Grant et al. 1976; Hook and Siraganian 1977). Furthermore, deuterium oxide (D_2O), which supports microtubule formation, appears to affect histamine release temperature dependently: at 37°C promotion prevails (Grant et al. 1977), whereas at lower temperature (25–32°C) inhibition becomes apparent (Hook and Siraganian 1977). Cytochalasin B, which disrupts microfilaments, markedly enhances liberation (Grant et al. 1977; Hook and Siraganian 1977). These findings suggest that the actual functional state of the microtubule system may affect peptide-induced release differently, whereas increases of the intracellular cAMP levels and intact microfilaments unambiguously hinder the reaction.

C5 peptide-induced histamine release from human basophils is reduced by the nonspecific serine esterase and phospholipase A_2 inhibitor bromophenacyl bromide, several lipoxygenase inhibitors, and by indomethacin. In contrast to the other inhibitors, indomethacin does not affect anti-IgE-elicited secretion (Farnam et al.1985). In accordance with IgE- and A 23187-induced release (Magro 1982), these findings suggest that lipoxygenase products – and less importantly, prostanoids – participate in peptide-elicited liberation.

2.6 Smooth Muscle and Heart Muscle Cells

Complement peptides contract several smooth muscle organs, namely blood vessels and tissues of the gastrointestinal and respiratory tracts. However, their characteristics differ profoundly from those of other spasmogens such as acetylcholine (ACh) and histamine: the peptides are generated de novo (no storage as preformed molecules), they induce rapid deactivation (tachyphylaxis) of target cells/organs and evoke release or formation of a flock of secondary mediators. The latter are often spasmogenically active themselves and may hence modulate or even cause the contractile response. Among these mediators, histamine, cyclooxygenase, and lipoxygenase products have been detected, but their actual significance for the contractile responses is often not fully understood. Beside such indirect stimulation, direct actions have conclusively been demonstrated for C5a on toad stomach cells only (Scheid et al. 1983).

2.6.1 Smooth Muscles of the Gastrointestinal Tract

In the gastrointestinal tract, C3a and C5a elicit contractile responses with pronounced species and organ dependence. Guinea pig intestines are much more sensitive than those from rat, rabbit, and hamster. Among guinea pig organs, the terminal ileum and uterus are most responsive (Friedberg et al. 1964; Kleine et al. 1970). Strips of isolated terminal ileum have therefore been used for many years to determine the biologic activity of complement peptides, and this has become the standard bioassay.

Not all complement peptides with other biologic activities are spasmogenic. This effect is exhibited predominantly by the primary peptides C3a, C4a, and C5a from all mammalian species studied so far, e.g., man, pig, guinea pig, and rat (reviewed by Hugli and Müller-Eberhard 1978). Furthermore, C5a-desArg from pig, guinea pig, and rat is active, whereas human C5a-desArg is largely devoid of activity and C3a-/C4a-desArg from all species tested is completely so. The spasmogenic activity of human C5a-desArg is apparently blocked by its bulky carbohydrate portion which presumably disturbs or masks the receptor-activating domain (but not in C5a). Enzymatic removal of this portion augments the spasmogenic potency about ten fold, but still amounting to only 1% of the activity of C5a (Gerard and Hugli 1981). For comparison, porcine C5a-desArg, which does not contain carbohydrates, is distinctly spasmogenic, being about half as potent as C5a (Damerau et al. 1980b; Gerard and Hugli 1981).

The high sensitivity of the guinea pig ileum is documented by the low threshold levels: they are about 10^{-10} M for human and porcine C5a, less than 10^{-9} M for porcine C5a-desArg and about 2×10^{-8} M for C3a from either species (Vallota and Müller-Eberhard 1973; Damerau et al. 1980b; Gerard and Hugli 1981; Hugli et al. 1981b). By contrast, human C4a is considerably less potent with a threshold at 10^{-6} M; it apparently binds to the C3a receptor. Its activity is abrogated by removal of the C terminal arginyl residue (Gorski et al. 1979).

In isolated guinea pig ileum, the peptides induce rapid, self-limiting contractions reaching a maximal strength similar to those of ACh or histamine. Because of their short duration, they can be clearly distinguished from the protracted responses to prostaglandins and peptide-containing leukotrienes LTC_4, LTD_4, and LTE_4. Contraction is usually produced by C3a only once (occasionally also up to 2–3 times), but

C5 peptides produce several contractions. Responsiveness to C5a and C5a-desArg fades from application to application due to rapidly developing deactivation (tachyphylaxis) (Friedberg et al. 1964, Vogt et al. 1969; Damerau et al. 1985b).

Contractile responses and the loss of responsiveness are largely separate processes. Both are initiated by the binding of the agents to specific cellular receptors. Then, the spasmogenic action of C5 peptides is mediated in isolated guinea pig ileum at least partly by histamine liberated from subserosal mast cells (Bodammer and Vogt 1970a; Sorgenfrei et al. 1982), because C5a-desArg degranulates mast cells at low concentrations (Damerau et al. 1985b), produces histamine release sufficient for contractile effects, and is inhibited by H_1 antihistaminics (Sorgenfrei et al. 1982). Additional mediation by Ach, serotonin, or PAF is unlikely (Vogt et al. 1969; Stimler et al. 1981a), part of the effect may be direct (Bodammer and Vogt 1970a). On the other hand, the action of porcine C3a seems to be virtually histamine independent. Earlier findings showing mast cell degranulation by C3a are not really contradictory since they were obtained with other organs and with C3a concentrations which considerably exceeded spasmogenic doses (discussed by Sorgenfrei et al. 1982).

Deactivation can be clearly distinguished from the contractile response: it develops under conditions (16°C, Ca^{2+}-free medium), which block contraction and histamine release (Damerau et al. 1985b). In accordance with earlier conclusions (Friedberg et al. 1964; Bodammer and Vogt 1970a), it is therefore not due to the exhaustion of histamine stores. Since deactivation is specific for C3 and C5 peptides and develops at low temperatures, it seems not to involve any post-receptor process. Rather, it is due to occupation of peptide receptors only: with repeated applications of C5a-desArg, loss of responsiveness strongly correlates with the cumulative contact times and follows first-order kinetics (Damerau et al. 1985b).

Deactivation is followed by recovery of the ileum strips within 20–30 min at 37°C. Recovery is highly temperature sensitive, blocked by the lysosomotropic agent chloroquine and by antimicrotubule agents such as cholchicine and vinblastine, but not affected by cycloheximide and cytochalasin B (Damerau et al. 1985c). These findings suggest that recovery proceeds via intralysosomal processing of the occupied receptors, which are regenerated and then incorporated into the plasma membrane by microtubule-dependent and microfilament-independent mechanisms.

Beside the specific, slowly reversible deactivation, a second type of unspecific deactivation has recently been discovered. This cross-deactivation is elicited by C3 and C5 peptides, and not by Ach or histamine.

Unlike specific deactivation, it depends on the contractile response of the ileum. It is maximal after a few seconds and fades within 3 min. Increases of intracellular cAMP in the direct target cells of C3a and C5a (subserosal mast cells?) may be causative as agents increasing cAMP levels promote whereas substances blocking cAMP elevations prevent it (Damerau et al. 1985d). This type of deactivation may serve as a negative feedback mechanism, which restrains stimulation of guinea pig ileum by complement peptides.

In contrast to the detailed studies on the guinea pig ileum, data on other intestinal organs are sparse. Rat ileum, which does not respond to histamine, is considerably less sensitive to C5 peptides. Inhibitor experiments indicate that neither histamine nor serotonin acts as a secondary mediator (Vogt et al. 1969). In the rat stomach, the spasmogenic effect of C5a-desArg (used as unfractionated activated rat serum) is inhibited by indomethacin. Therefore, cyclooxygenase products may be involved, whereas histamine, serotonin, and bradykinin do not mediate the contractile response (Damas 1982).

2.6.2 Smooth Muscles of Lung and Trachea

In recent years, actions of complement peptides on the respiratory system have regained considerable interest going far beyond the earlier analyses of shock-promoting effects in guinea pigs. Though performed with animal tissues, recent studies center on pathomechanisms of human pulmonary diseases (bronchial asthma, lung inflammations, and ARDS).

In strip-like preparations of guinea pig trachea and lung, C3a and C5 peptides induce slowly increasing, prolonged contractions which become maximal after some minutes and fade within 10–40 min (Regal et al. 1980; Stimler et al. 1981b). In lung strips the contractions can be completely removed by washing (Stimler et al. 1981b). Sensitivity of both organs to the peptides is similar to that of the guinea pig ileum (see previous section). However, in contrast to its behavior in the ileum, in lung preparations C3a has much lower intrinsic activity than C5 peptides.

Upon repeated applications, both lung and trachea are rapidly rendered unresponsive to C5a (Regal et al. 1980; Stimler et al. 1981b, 1982). Selective deactivation to C3 and C5 peptides has been observed in lung tissue (Stimler et al. 1981b). In contrast, the synthetic C terminal fragment C3a(57–77), which is as potent as native C3a, does not cause deactivation, possibly due to different binding characteristics (Huey et al. 1984b).

Pharmacologic experiments suggest that the peptides act via various secondary mediators which are as yet incompletely defined. Interpretation of the data is difficult, since the action of some mediators seems to be based on cooperative processes. Furthermore, for most inhibitors unspecific effects have not been thoroughly ruled out.

Histamine appears to be of minor importance, since H_1 antihistaminics are largely ineffective, except for inhibition of the onset of C5a-induced contractions of lung strips (Regal et al. 1980, 1983b; Stimler et al. 1981b, 1982, 1983). In lung preparations, SRS-A (slow reacting substance of anaphylaxis) may predominantly mediate the effect of C5a. In addition, endogenous PGE_1 and $PGF_{1\alpha}$ may either be involved directly in the contractile response (Stimler et al. 1982; Regal et al. 1983b) or may facilitate the action of SRS-A, since cyclooxygenase inhibitors impair the effects of LTC_4 and LTD_4 (Stimler 1984). Beside such functional evidence for their involvement, formation/release of SRS-A, prostaglandins, and histamine upon stimulation with C5 peptides has been reported (Stimler et al. 1982, 1983). The C3a-induced response, on the other hand, seems to be predominantly mediated by cyclooxygenase products, because it is effectively reduced by aspirin and indomethacin, but not by lipoxygenase inhibitors. Accordingly, C3a stimulates release of PGE_2 and $PGF_{1\alpha}$, but not of SRS-A (Stimler et al. 1983).

In tracheal preparations, aspirin does not affect, but indomethacin moderately curtails the response to C5a, whereas in combination with the H_1 antihistaminic diphenhydramine both drugs evoke distinct inhibition. Therefore, Regal and Pickering (1983) have assumed that histamine and cyccloxygenase products evoke the response in mutual cooperation. However, PAF could also act as a secondary mediator in this response: it is formed in various cell types after stimulation with complement peptides, causes bronchoconstriction in guinea pigs, and is inhibited by combinations of the aforementioned drugs (Vargaftig et al. 1982). In addition to these putative mediators, leukotrienes (SRS-A) may be involved (Regal and Pickering 1981).

2.6.3 Blood Vessels and Heart Muscle

In the cardiovascular system, complement peptides evoke an array of pathologic changes which may aggravate disturbances in shock states by occlusion of blood vessels and by deranging cardiac functions (see Sect. 3.2). These alterations are produced mainly by actions on vascular smooth muscle cells and heart muscle cells, as well as by intravascular formation of leukocyte and, in certain species, of platelet aggregates.

In the guinea pig heart, C3a and C5a cause positive inotropic and chronotropic effects in spontaneously beating atria in Langendorff-perfused hearts (Greef et al. 1959; Huey et al. 1984a; Hachfeld del Balzo et al. 1985; Regal 1985) and in electrically stimulated papillary muscle

(Bernauer et al. 1971, 1972). By contrast, rat papillary muscle fails to respond to C5a-desArg (Bernauer et al. 1972). Quantitatively, the positive inotropic action on guinea pig atria is well in the physiologic range: thresholds are 10^{-9} M for C5a and 4×10^{-9} M for C3a, the latter being surprisingly potent in comparison with other actions (Huey et al. 1984a).

At low stimulus intensity, repeated applications seem not to induce deactivation (Huey et al. 1984a), but at higher concentrations C5 peptides cause rapid loss of reactivity (Regal 1985). The chronotropic response is partly inhibited by the H_2 antihistaminics metiamide and cimetidine, indicating that endogenous histamine mediates the effect at least initially. In contrast, the action of C3a on atria is virtually histamine-independent (Huey et al. 1984a). However, in Langendorff-perfused heart preparations histamine is liberated by C3a and its amount closely correlates with the positive chronotropic changes (Hachfeld del Balzo et al. 1985). Therefore, histamine seems to originate mainly from other cardiac regions. In atria, FPL 55712 inhibits both peptides, whereas indomethacin impedes the effect of C3a only, but even slightly augments that of C5a (Huey et al. 1984a). Therefore, the authors conclude that C3a responses are mediated by leukotrienes and prostaglandins, those to C5a by leukotrienes and vasoactive amines.

As indicated before, the effects on isolated Langendorff-perfused hearts are complex. Unlike atria, the ventricles respond with a decrease of contractile force to C3a and C5a (Bernauer et al. 1971; Hachfeld del Balzo et al. 1985). Furthermore, the peptides stimulate sinus rate, impair atrioventricular conduction, and induce coronary vasoconstriction. C3a-desArg is inactive. The negative inotropic effect may be caused by endogenous leukotrienes; they appear not to be due to the concomitant coronary vasoconstriction. The coronary artieres are presumably constricted by endogenous thromboxane, which has been detected in measurable amounts after stimulation with C3a (Hachfeld del Balzo et al. 1985).

Regarding the pathophysiologic implications of these findings, Hachfeld del Balzo et al. (1985) suggest that in immediate hypersensitivity reactions complement peptides may contribute to cardiac dysfunctions such as tachycardia, arrhythmia, left ventricular contractile failure, and coronary vasoconstriction. This seems to be feasible – at least for disturbances mediated by histamine – since human hearts contain relatively large amounts of this amine (for a survey see Levi et al. 1982).

In vivo, systemic application of C5a-desArg causes a short fall and subsequently a longer-lasting rise of peripheral arterial blood pressure in guinea pigs; in cats, only the hypotensive reaction is inducible (Bodammer 1969).

Similarly, infusion of human and porcine C5a into the renal artery of rats leads to increases in efferent arteriolar resistance resulting in decreased nephron plasma flow (Pelayo et al. 1986). Experiments with strips of isolated blood vessels have shown that the reactions to the peptides are highly species dependent: whereas strips from guinea pig vessels are contracted by C3a and C5a with thresholds of 10^{-9} M and 10^{-8} M (Regal 1982; Marceau and Hugli 1984), preparations from rabbits respond to C5a predominantly with relaxation (threshold about 10^{-9} M) and are resistant to C3a (Hugli and Marceau 1985). In both animal species, the sensitivity decreases from portal vein to pulmonary artery to aorta. Contractile responses to C3a or C5a last much shorter than those to other vascoconstrictors such as norepinephrine and histamine. With repeated applications deactivation develops rapidly. Analyses with various inhibitors indicate that the effects of C3a and C5a on vascular tissues are mediated by different endogenous agents. The responses of guinea pig vessel preparations seem to be evoked by prostaglandin-like substances (TxA_2 ?), with additional participation of leukotrienes in the case of C3a and of histamine in the case of C5a (Regal 1982; Marceau and Hugli 1984). The relaxing effect of C5a on rabbit vessels may be due to the vasodilatory prostaglandins PGE_2 and PGI_2 (Hugli and Marceau 1985). Because of their short duration and the rapid deactivation, vascular actions of complement peptides seem to be of minor physiologic significance.

3 Actions In Vivo

3.1 Proinflammatory Effects

When applied locally to extravascular sites in vivo, complement peptides produce acute inflammatory reactions with increases in vascular permeability and accumulation of leukocytes. The models used range from injections into solid tissues (intradermal/subcutaneous sites) and body cavities (peritoneal/pleural cavity) to instillations into the respiratory tract (for references see below). Proinflammatory effects have been demonstrated with purified peptides in different mammals such as man, rabbit, guinea pig, and hamster.

Depending on animal species, kind of peptide, and site of application, the data show some divergences, especially with respect to those on leukocyte accumulation. This effect is usually elicited by human, porcine, and guinea pig C5 peptides (Jensen et al. 1969; Damerau and Vogt 1976; Fernandez et al. 1978; Larsen et al. 1980; Björk et al. 1985;

Yancey et al. 1985), but could not be shown with human C3a (Fernandez et al. 1978; Björk et al. 1985). In contrast to the human C3 peptide, porcine C3a and C3a-desArg induce leukocyte accumulation (Damerau et al. 1978b). The same species difference is seen in an in vitro model (Sykes-Moore chamber): here human C3a-desArg fails to stimulate undirected migration of human PMN, while porcine C3a-desArg is active (own unpublished observations). Accordingly, porcine C3a and C3a-desArg but not human C3a evoke PMN chemotaxis (Damerau et al. 1978a; Fernandez et al. 1978). An example of the influence of the application site is the finding that intratracheal instillation in rabbits of human C5a shows no or only weak proinflammatory action, albeit that C5a-desArg causes severe inflammation (Larsen et al. 1980; Stimler et al. 1980). Since in vitro (chemotaxis, exocytosis) C5a is more active than its desArg derivative, the authors suggest that in vivo the peptides may bind to different structures/cells or may be distributed differently. For instance, primary C3a and C5a could be eliminated more rapidly from the site of application, because they are more potent histamine releasers. Furthermore, different cell types may respond differently to C5a and C5a-desArg, or kinds/amounts of mediators released from these cells may vary and lead to different synergistic effects (Huey et al. 1983).

As a further speciality, vasodilators are sometimes required as additional mediators of proinflammatory actions. This synergism with complement peptides and other chemoattractants (LTB_4, FMLP) seems to be restricted to tissues with generally little blood supply, e.g., skin (Wedmore and Williams 1981), but may not be required in regions with higher perfusion rates, such as the guinea pig pleural wall or hamster cheek pouch. Alternatively, in other regions endogenous vasodilators may be produced upon stimulation with complement peptides and thus facilitate the proinflammatory effects. For example, increased production of PGI_2 by C5a and C5a-desArg has been demonstrated in isolated aortic strips from rabbits (Rampart et al. 1983).

In view of such special requirements, it is not surprising that proinflammatory potencies of complement peptides vary considerably in the different studies. Intradermal and subcutaneous injections seem to be most effective: permeability increases in human or guinea pig skin are evoked by doses as low as 10 pg – 1 ng C5a, 100 ng porcine C5a-desArg and 1 μg human C5a-desArg (Vallota and Müller-Eberhard 1973; Hugli et al. 1981b; Stimler et al. 1981a; Yancey et al. 1985). In contrast, human C5a-desArg has been reported to be without effect in guinea pig skin (Webster et al. 1980a). C3a is active in the nanogram, C4a in the microgram range (Wuepper et al. 1972; Vallota and Müller-Eberhard

1973; Gorski et al. 1979; Stimler et al. 1981a). Astonishingly, in the hamster cheek pouch C3a is about ten-fold as active as C5a (Björk et al. 1985). When applied into body cavities, 100–1000 times higher peptide doses are required to increase permeability (Damerau and Vogt 1976; Damerau et al. 1978b; Fernandez et al. 1978). Comparing the leukocyte-accumulating potencies in rabbit skin on a molar basis, C5a-desArg has been found to be markedly more active than LTB_4 and PAF, but distinctly less potent than *Escherichia coli* endotoxin (Colditz and Movat 1984b; Movat et al. 1984).

A close correlation between leukocyte-accumulating and permeability-increasing effects has been observed for porcine C3 peptides as well as human and porcine C5 peptides (Damerau and Vogt 1976; Damerau et al. 1978b; Wedmore and Williams 1981; Björk et al. 1985). This points to a causal relationship between both phenomena. Indeed, in certain systems complement peptides produce vascular leakage only in the presence of PMN. In leukopenic animals the peptides fail to augment permeability, while low molecular weight agents (histamine, bradykinin) exhibit direct, PMN-independent activity (Issekutz et al. 1980; Wedmore and Williams 1981).

Mechanistically, PMN have been proposed to break junctions between adjacent endothelial cells and to damage the underlying basement membrane by their lysosomal proteases or cytotoxic oxygen species. Alternatively, stimulated PMN may secrete substances which cause contraction of endothelial cells and thus lead to permebility increases (Issekutz et al. 1980; Wedmore and Williams 1981). A major role for PMN-derived H_2O_2 in endothelial damage is suggested by the inhibitory action of catalase (Rampart and Williams 1985). On the other hand, neither endogenous LTB_4 nor histamine appears to cooperate in the PMN-dependent effect of C3a and C5a, whereas elsewhere histamine may contribute to early leakage (Bodammer and Vogt 1970b; Forrest et al. 1981; Björk et al. 1985; Yancey et al. 1985).

In addition to their PMN-dependent action, complement peptides may also increase permeability directly: porcine C3a-desArg has been found to affect vascular walls even when PMN accumulation is drastically reduced by concomitant systemic application of this peptide or by treatment with colchicine. As shown by inhibitor experiments, C3a-desArg acts independently of histamine and prostanoids (Damerau et al. 1978b).

Further insights into the regulation of proinflammatory activities were gained by studies on lung inflammation. Intravenous injection of C5a, though provoking massive PMN margination and trapping of aggregates in lung vessels, does not create inflammatory lesions (Henson et al. 1979). Actual inflammatory response with increased permeability, fibrin formation, diapedesis of erythrocytes and sequential accumulation of PMN and monocytes is elicited only when the peptide is instilled extravascularly, into the airways (Kreutzer et al. 1979; Shaw et al. 1980). Therefore, additonal factors are apparently needed for its initiation. One of these may be a concentration gradient originating from the tissue space, by which the chemoattractant could attract PMN into the lesion. Specifically for the lung, alveolar macrophages have been demonstrated as releasing a PMN-attracting agent upon stimulation with C5 peptides (McCarthy and Henson 1979). Furthermore, PMN may damage endothelial cells during their transvascular migration only, for instance by generating oxygen radicals (Sacks et al. 1978). Another facilitating factor may be the formation of prostanoids after "mild" insults to the lung (surgical procedures, transient hypoxia), as they work synergistically with C5 peptides (Larsen et al. 1985).

In contrast to the advancing analysis of additional factors, the mechanisms regulating intensity and duration of inflammatory reactions are just beginning to be understood. Until recently, it had been generally assumed that proinflammatory actions of complement peptides were terminated predominantly by (proteolytic) inactivation, receptor binding, and deactivation of the effector PMN (Goldstein 1979; Meuer et al. 1982). However, these mechanisms may be operative in the blood stream only. Here complement peptides actually diminish PMN responsiveness resulting in impaired emigration (Damerau et al. 1978b; Solomkin et al. 1981). Recent studies in which chemoattractants were repeatedly applied into rabbit skin suggest other local regulatory mechanisms: here, PMN emigration appears not to be solely controlled by concentration gradients (Colditz and Movat 1984b). In addition, it is a self-limiting process, but it is limited neither by deactivation of circulating PMN or by generation of unspecific migration inhibitors (Colditz and Movat 1984a). Rather, it is terminated by deactivation of other targets, presumably endothelial cells or cells remote from the vessel wall. Since deactivation in this model is essentially stimulus-specific (with partial cross-deactivation between LTB_4, C5 fragments, and PAF), Colditz and Movat (1984a) conclude that PMN accumulation is terminated by downregulation of a receptor-coupled mechanism which initially triggers the emigration response.

In natural inflammatory processes, several PMN activators and additional amplification mechanisms, resulting in network-like multistep reactions, are likely to work simultaneously, with different compositions of agents and cells. It is, therefore, very difficult to evaluate the pathophysiologic significance of C3a or C5a for individual diseases. Nevertheless, complement peptides have been detected, partly in concentrations sufficient to evoke biologic actions, in various inflammatory fluids/ lesions, such as in joint fluids from rheumatoid arthritis, gouty arthritis, vasculitis-affected tissue, peritoneal exudates and cerebrospinal fluid during meningitis (reviewed by Vogt 1974, 1986; Hugli and Müller-Eberhard 1978). Hence, complement peptides most likely contribute to symptoms in these diseases, but their actual significance is not known in most cases.

3.2 Shock-Promoting Effects

The original denomination of porcine, guinea pig and rat C5a-desArg as anaphylatoxin points to its acute life-threatening effects, most severe in guinea pigs, after systemic application. In the guinea pig the main symptoms are bronchospasm and cardiovascular disturbances. Intravenous bolus injection of C5a-desArg causes a short-lasting tachypnoea which does not undergo deactivation at repeated challenges. It is, however, abolished by bilateral vagotomy, indicating that this response is produced by reflex mechanisms (Bodammer and Vogt 1967; Bodammer 1968). The initial stimulatory phase is soon followed by severe impairment of expiration leading to lung inflation and eventually to death. The underlying bronchospasm is reminiscent of that in human asthmatic attack. It seems to be induced by the combined spasmogenic actions of liberated histamine, serotonin, and of cholinergic stimulation since it is partly inhibited by each tripelennamine, methysergide, and atropine. Combinations of tripelennamine and atropine totally abolish it (Bodammer and Vogt 1967; Bodammer 1968). In cats, tachypnoea and bronchospasm are less pronounced; they seem to be mediated predominantly by histamine (Bodammer 1969).

In guinea pigs, the changes in blood pressure are less severe than, and independent from the respiratory disturbances: systemic administration of porcine C5a-desArg produces first a transient fall (not subject to deactivation) and then a rise in peripheral arterial blood pressure, which rapidly becomes tachyphylactic upon restimulation. Both alterations disappear spontaneously after a few minutes. The initial hypotensive reaction is not mediated by histamine, serotonin, ACh, or catecholamines (Bodammer and Vogt 1967). Actually, it may be due to production of vasodilatory PGI_2 and to obstruction of lung vessels by platelet and leukocyte aggregates (Grossklaus et al. 1976; Stimler et al. 1980). The latter changes presumably elevate the pulmonary artery pressure (determined in cats, but for technical reasons not in guinea pigs; Bodammer 1969) and decrease cardiac output. As has been shown in sheep (McDonald et al. 1983), formation of TxA_2 appears to contribute to the increased resistance in pulmonary circulation. In guinea pigs, the hypertensive response to C5a-desArg is presumably elicited by catecholamines as it is blocked by dihydroergotamine (Bodammer and Vogt 1967). In cats, both depressor and pressor response become rapidly tachyphylactic; they seem to be independent of histamine, serotonin, ACh, catecholamines, and prostanoids. Since the depressor response clearly predominates in this species (Bodammer 1969), the circulatory changes may be created by platelet and leukocyte aggregates obstructing pulmonary circulation. Similar symptoms, which are most likely due to the same pathomechanisms, have been observed in dogs and sheep (Pavek et al. 1979).

In humans, complement peptides may also produce similar cardiovascular disturbances. Their generation in the blood stream, which would preferably trigger such events, has been demonstrated, for instance, during hemodialysis. leukapheresis, cardiopulmonary bypass, ARDS, severe trauma, sepsis, and burn injury (Craddock et al. 1977a; Hammerschmidt et al. 1978a; Chenoweth et al. 1981; Solomkin et al. 1981; Heideman and Hugli 1984; review by Vogt 1986). Conversely, anti-C5a antibodies have been found to mitigate the respiratory distress elicited in primates (*Macaca fascicularis*) by intravenous infusion of *E. coli* (Stevens et al. 1986). Earlier, it had been assumed that major shock symptoms developing in these instances could be elicited directly by peptide-induced PMN stimulation, in particular by aggregate formation and subsequent occlusion of small blood vessels (Hammerschmidt et al. 1978b; Jacob et al. 1980; Jacob 1981). However, the role of PMN aggregates had presumably been overestimated, since they disintegrate spontaneously in vivo. Therefore, PMN aggregation must be accompanied by other pathologic changes in order to disturb the microcirculation seriously. Some of these – largely putative – mechanisms have been discussed in previous chapters (hypoxemia, formation of TxA_2; see for instance Gee et al. 1985); additionally, for sustained vascular damages platelet activation (via PMN-derived PAF?) (Tvedten et al. 1985), secondary local activation of the coagulation system (Muhfelder et al. 1979), priming of PMN for increased exocytosis (Fowler et al. 1984), prolonged complement activation, and formation of oxygen radicals (Till et al. 1982; Ward et al. 1985), as well as synergistic actions between complement peptides and bacterial lipopolysaccharides on PMN (Smedley et al. 1986) may be required.

4 Conclusions

The complement-derived peptides C3a, C4a, C5a, and their desArg derivatives are generated in vivo when either foreign materials/surfaces (bacteria, viruses, plastic/cellulose compounds) encounter the organism and activate the complement system, or when, after tissue damage cellular proteases are liberated and directly split them off from their parent components. C5 peptides – and to a lesser extent C3 peptides – can be generated in vivo in sufficient amounts to exert biologic effects (about 1% of the maximally attainable peptide concentration can be sufficient).

When locally formed in tissues, complement peptides usually act in their vicinity, predominantly as inflammatory mediators which activate PMN and increase vascular permeability. There, they may initiate/support beneficial defense reactions or may help to damage host tissues during self-perpetuating, chronic inflammatory processes. On the other hand, intravascular formation seems to be detrimental to the organism, namely by induction of cell aggregation, occlusion of blood vessels, endothelial damage, and cardiovascular disturbances, e.g., by release of histamine and formation of vascoconstricting secondary mediators such as TxA_2.

As regards the biologic significance of the individual peptides, only C5a-desArg is a species-independent effector molecule, having a much better stability in vivo than C5a and high biologic potency in comparison with that of the C3 peptides. C3a and C3a-desArg may participate under certain conditions in inflammatory processes and shock-like events. However, the sensitivities of the organisms to and the potencies of the C3 peptides vary considerably from species to species. Furthermore, both peptides usually have a rather low potency relative to the concentrations attainable in vivo. For C4a and C4a-desArg no significant role can be envisaged at present since they are inactive or of low activity. In general, the primary peptides (C3a, etc.) can be assumed to be of minor relevance, as their life time in vivo is in the range of seconds only before they are attacked by ubiquitiously occurring carboxypeptidases. However, in certain microenvironments in which these enzymes may be absent or their activity is possibly blocked (such as in the acidic milieu in inflamed tissues) the primary peptides may play a more important role.

The phlogistic effects of C5 peptides – and of C3 peptides from certain species – lead to acute inflammatory processes, including the whole array of symptoms, and possibly triggering various effector systems. Of the latter, their action on PMN is best documented and seems to be most important. Physiologically, the peptides cause adhesion and stimulate migration (chemotaxis) of PMN. Other induced PMN actions, such as exocytosis and formation of oxygen-containing radicals, have been demonstrated mainly in vitro, but their in vivo role cannot presently be evaluated; strong adhesion, which may not be reached in the organism, is indispensable for these responses.

Vascular permeability is enhanced by all complement peptides, predominantly by C5a-desArg which on a molar basis, is more active than amines. This response is caused partly by direct effects on the vascular wall and partly by indirect actions involving activated marginating PMN and vasodilators such as PGI_2 and PGE_1. The respective interactions of these mediators vary with animal species and site of application.

Less certainly, complement peptides (C5-desArg) may propagate and regulate inflammatory processes by stimulating monocytes/macrophages and mast cells/basophils to release their granule constituents and to move. These actions point to a possible involvement of these agents in chronic inflammation and in acute allergic reactions. However, strict evidence for such functions is still missing.

Of eminent biologic interest, albeit with the least ascertained observations in vivo, are the antagonistic actions of C3a (inhibitory) and of both C5 peptides (stimulatory) on primary immune responses. The largely speculative, yet fascinating consequence of these in vitro findings lies in the assumption of an important regulatory function of complement peptides, namely shifting unspecific to specific, immunologic defense reactions and vice versa: in complement activations with C3, but not C5 conversion (such as in the fluid phase by soluble activators which may usually be eliminated rapidly) unspecific inflammatory reactions with predominant vascular responses but with minor PMN involvement and without specific antibody production should develop. On the other hand, complete complement activation, including C5 cleavage which is induced by particulate materials exhibiting appropriate surface properties (for instance, certain carbohydrates or antibodies), would result in inflammatory reactions with stimulation of PMN and monocytes/macrophages as well as triggering of the primary immune response as the second line of defense in the body.

Further in vitro actions such as platelet activation and smooth muscle contraction are, in my opinion, of minor pathophysiologic relevance for the human organism, but seem to be more important in sensitive animal species such as in guinea pigs. In man, secondary desArg peptides may contribute to certain disease states (e.g., bronchial asthma, hemostatic disorders) by pathologic formation of secondary effector molecules such as leukotrienes and TxA_2, but only when the underlying disease leads to altered responsiveness of target cells and/or enhanced mediator formation.

Beside the peptides' potencies and the responsiveness of individual target cells/organs, the binding properties of complement peptides (fast and strong binding to specific receptors) as well as the often rapid loss of cellular reactivity (due to receptor down-regulation, which may already develop at subthreshold concentrations) must be taken into account. These features curtail those reactions requiring rapid cell stimulation (such as aggregatory and secretory responses in PMN, mast cells/basophils, platelets, and contractions of certain smooth muscle cells) and may thereby prevent unnecessary and excessive responses. By contrast, PMN/endothelial interaction, cell locomotion, and increases of vascular permeability are much less prone to such negative feedback regulation. Therefore, they can be triggered and proceed during slow and prolonged generation of complement peptides.

Acknowledgment. I wish to sincerely thank Miss Y. Haller for typing the long list of references.

References

Babior BM (1984) The respiratory burst of phagocytes. J Clin Invest 73:599–601

Baggiolini M, Dewald B (1985) The neutrophil. Int Arch Allergy Appl Immunol [Suppl 1] 76:13–20

Bass DA, Gonwa TA, Szejda P, Cousart MS, DeChatelet LR, McCall CE (1980) Eosinopenia of acute infection. Production of eosinopenia by chemotactic factors of acute infection. J Clin Invest 65:1265–1271

Baumgarten H, Zierz R, Schulze M, Götze O (1985) Identification and quantification of the fifth component of complement (C5) as a membranemarker for human monocytes. Complement 2:7–8

Becker EL, Henson PM (1973) In vitro studies of immunologically induced secretion of mediators from cells and related phenomena. Adv Immunol 17:93–193

Becker EL, Showell HJ (1972) The effect of Ca^{2+} and Mg^{2+} on the chemotactic responsiveness and spontaneous motility of rabbit polymorphonuclear leukocytes. Z Immunitatsforsch 143:466–476

Becker EL, Showell HJ, Henson PM, Hsu LS (1974) The ability of chemotactic factors to induce lysosomal enzyme release. I. The characteristics of the release, the importance of surfaces and the relation of enzyme release to chemotactic responsiveness. J Immunol 112:2047–2054

Becker EL, Sigman M, Oliver JM (1979) Superoxide production induced in rabbit polymorphonuclear leukocytes by synthetic chemotactic peptides and A 23187. The nature of the receptor and the requirement for Ca^{2+}. Am J Pathol 95: 81–97

Becker S, Hadding U, Schorlemmer HU, Bitter-Suermann D (1978a) Demonstration of high-affinity binding sites for C3a anaphylatoxin on guinea-pig platelets. Scand J Immunol 8:551–555

Becker S, Meuer S, Hadding U, Bitter-Suermann D (1978b) Platelet activation: a new biological activity of guinea-pig C3a anaphylatoxin. Scand J Immunol 7:173–180

Bender JG, McPhail LC, van Epps DE (1983) Exposure of human neutrophils to chemotactic factors potentiates activation of the respiratory burst enzyme. J Immunol 130:2316–2323
Benner KU, Schumacher KA, Classen HG (1975) Platelet aggregation induced by anaphylatoxin and its inhibition. Drug Res 25:1635–1638
Bernauer W, Hahn F, Wissler J, Nimptsch P, Filipowski (1971) Untersuchungen mit klassischem und kristallisiertem Anaphylatoxin an isolierten Herzpräparaten. Naunyn Schmiedebergs Arch Pharmakol 269:413
Bernauer W, Hahn F, Nimptsch P, Wissler J (1972) Studies on heart anaphylaxis. V. Cross-desensitization between antigen, anaphylatoxin and compound 48/80 in the guinea pig papillary muscle. Int Arch Allergy Appl Immunol 42:136–151
Björk J, Hugli TE, Smedegard G (1985) Microvascular effects of anaphylatoxins C3a and C5a. J Immunol 134:1115–1119
Bodammer G (1968) Untersuchungen zum Mechanismus der Atemwirkung des Anaphylatoxins. Naunyn Schmiedebergs Arch Pharmakol Exp Pathol 260: 16–25
Bodammer G (1969) Untersuchungen über den Mechanismus der Blutdruckwirkung des Anaphylatoxins bei Katzen und Meerschweinchen. Naunyn Schmiedebergs Arch Pharmakol Exp Pathol 262:197–207
Bodammer G, Vogt W (1967) Actions of anaphylatoxin on circulation and respiration of the guinea pig. Int Arch Allergy Appl Immunol 32:417–428
Bodammer G, Vogt W (1970a) Contraction of the guinea pig ileum induced by anaphylatoxin independent of histamine release. Int Arch Allergy Appl Immunol 39:648–657
Bodammer G, Vogt W (1970b) Beeinflussung der Capillarpermeabilität in der Meerschweinchenhaut durch Anaphylatoxin (AT). Naunyn Schmiedebergs Arch Pharmacol 266:255–260
Boyden S (1962) Chemotactic effect of mixtures of antibody and antigen on polymorphonuclear leukocytes. J Exp Med 115:453–466
Camussi G, Tetta C, Bussolino F, Cappio FC, Coda R, Masera C, Segolini G (1980) Mediators of immune complex-induced aggregation of polymorphonuclear neutrophils. I. C5a anaphylatoxin, neutrophil cationic proteins and their cleavage fragments. Int Arch Allergy Appl Immunol 62:1–15
Camussi G, Tetta C, Bussolini F, Cappio FC, Masera C, Segolini G (1981a) Mediators of immune-complex-induced aggregation of polymorphonuclear neutrophils. II. Platelet-activating factor as the effector substance of immune-induced aggregation. Int Arch Allergy Appl Immunol 64:25–41
Camussi G, Tetta C, Segolini G, Deregibus MC, Bussolini F (1981b) Neutropenia induced by platelet-activating factor (PAF-acether) released from neutrophils: the inhibitory effect of prostacyclin (PGI_2). Agents Actions 11:550–553
Charriaut C, Senik A, Kolb JP, Barel M, Frade R (1982) Inhibition of in vitro natural killer cell activity by the third component of complement: role for the C3a fragment. Proc Natl Acad Sci USA 79:6003–6007
Chenoweth DE, Hugli TE (1978) Demonstration of specific C5a receptor on intact human polymorphonuclear leukocytes. Proc Natl Acad Sci USA 75:3943–3947
Chenoweth DE, Hugli TE (1980) Human C5a and C5a analogs as probes of the neutrophil C5a receptor. Mol Immunol 17:151–161
Chenoweth DE, Cooper SW, Hugli TE, Stewart RW, Blackstone EH, Kirklin JW (1981) Complement activation during cardiopulmonary bypass. Evidence for generation of C3a and C5a anaphylatoxins. N Engl J Med 304:497–503
Clancy RM, Dahinden CA, Hugli TE (1983) Arachidonate metabolism by human polymorphonuclear leukocytes stimulated by N-formyl-Met-Leu-Phe or complement component C5a is independent of phospholipase activation. Proc Natl Acad Sci USA 80:7200–7204

Cochrane CG, Müller-Eberhard HJ (1968) The derivation of two distinct anaphylatoxin activities from the third and fifth component of human complement. J Exp Med 127:371–386

Colditz IG, Movat HZ (1984a) Desensitization of acute inflammatory lesions to chemotaxins and endotoxin. J Immunol 133:2163–2168

Colditz IG, Movat HZ (1984b) Kinetics of neutrophil accumulation in acute inflammatory lesions induced by chemotaxins and chemotaxinigens. J Immunol 133:2169–2173

Craddock PR, Fehr J, Brigham KL, Kronenberg RS, Jacob HS (1977a) Complement and leukocyte-mediated pulmonary dysfunction in hemodialysis. N Engl J Med 296:769–774

Craddock PR, Fehr J, Dalmasso AP, Brigham KL, Jacob HS (1977b) Hemodialysis leukopenia. Pulmonary vascular leukostasis resulting from complement activation by dialyzer cellophane membranes. J Clin Invest 59:879–888

Craddock PR, Hammerschmidt D, White JG, Dalmasso AP, Jacob HS (1977c) Complement (C5a)-induced granulocyte aggregation in vitro. A possible mechanism of complement-mediated leukostasis and leukopenia. J Clin Invest 60:260–264

Craddock PR, White JG, Jacob HS (1978) Potentiation of complement (C5a)-induced granulocyte aggregation by cytochalasin B. J Lab Clin Med 91:490–499

Craddock PR, Hammerschmidt DE, Moldow CF, Yamada O, Jacob HS (1979) Granulocyte aggregation as a manifestation of membrane interactions with complement: possible role of leukocyte margination, microvascular occlusion, and endothelial damage. Sem in Hematol 16:140–147

Czarnetzki BM, Pawelzik B (1983) The effect of eosinophil chemotactic factors on in vitro eosinopoiesis in the guinea pig. Blut 47:1–10

Dahinden CA, Fehr J, Hugli TE (1983) Role of cell surface contact in the kinetics of superoxide production by granulocytes. J Clin Invest 72:113–121

Damas J (1982) An unusual prostaglandin-mediated response of rat stomach strip to anaphylatoxin. Arch Int Pharmacodyn Ther 256:163–165

Damerau B, Keller HU (1983) Comparison of adhesion to glass and aggregation of human leukocytes. Agents Actions 13:53–58

Damerau B, Vogt W (1976) Effect of hog anaphylatoxin (C5a) on vascular permeability and leukocyte emigration in vivo. Naunyn Schmiedebergs Arch Pharmacol 295:237–241

Damerau B, Vogt W (1982) Pseudo-allergic reactions based on effects of complement-derived peptides on circulating leukocytes. In: Dukor P, Kallos P, Schlumberger HD, West GB (eds) Involvement of drugs and chemicals. Karger, Basel, pp 101–121 (Pseudo-allergic reactions, vol 3)

Damerau B, Grünefeld E, Vogt W (1978a) Chemotactic effects of the complement-derived peptides C3a, $C3a_i$ and C5a (classical anaphylatoxin) on rabbit and guinea pig polymorphonuclear leukocytes. Naunyn Schmiedebergs Arch Pharmacol 305:181–184

Damerau B, Höllerhage HG, Vogt W (1978b) Effect of the cleavage peptides, C3a and $C3a_i$, from the third component of hog complement on leukocyte accumulation and vascular permeability in vivo. Naunyn Schmiedebergs Arch Pharmacol 302:45–50

Damerau B, Grünefeld E, Vogt W (1980a) Aggregation of leukocytes induced by the complement-derived peptides C3a and C5a and by three synthetic formylmethionyl peptides. Int Arch Allergy Appl Immunol 63:159–169

Damerau B, Zimmermann B, Grünefeld E, Czorniak K, Vogt W (1980b) Biological activities of C5a and C5a-desArg from hog serum. Int Arch Allergy Appl Immunol 63:408–414

Damerau B, Grünefeld E, Vogt W (1982) Influence of indometacin, phenylbutazone and BW 755C on different biological activities of C3a and C5a-desArg. Mol Immunol 19:1408

Damerau B, Wüstefeld H, Vogt W (1983) Binding characteristics of the complement peptides C3a and C5a-desArg to cellulose nitrate filters in Boyden chambers. Agents Actions [Suppl] 12:121–132

Damerau B, Wüstefeld H, Fricke D, Kunze H (1984) Influence of the anti-inflammatory serine esterase inhibitor gabexate mesilate (Foy) on aggregation, locomotion and adhesion of polymorphonuclear leukocytes. Agents Actions 15: 594–599

Damerau B, Haller Y, Vogt W (1985a) Influence of surface contact on complement peptide induced activities of human polymorphonuclear leukocytes (PMN). Complement 2:20

Damerau B, Roesler J, Vogt W (1985b) Loss and recovery of sensitivity of guinea pig isolated ileum to the spasmogenic action of the complement peptide C5a-desArg. Br J Pharmacol 84:47–54

Damerau B, Roesler J, Vogt W (1985c) Pharmacological characterization of the slow component of deactivation of guinea pig isolated ileum to the spasmogenic action of C5a-desArg. Br J Pharmacol 84:55–61

Damerau B, Roesler J, Vogt W (1985d) Fast deactivation of guinea pig isolated ileum to C5a-desArg: a possible cyclic AMP-dependent mechanism. Br. J Pharmacol 84:63–69

Damerau B, Zimmermann B, Czorniak K, Wüstefeld H, Vogt W (1986a) Role of the N-terminal regions of hog C3a, C5a and C5a-desArg for their biological activities. Mol Immunol 23:433–440

Damerau B, Otte G, Haller Y, Löffler BM (1986b) Einfluß von Benzydamin auf verschiedene Funktionen von Human-Granulozyten und auf deren Interaktion mit Endothelzellen. Drug Res (in press)

Daughaday CC, Mehta J, Spilberg I, Atkinson JP (1985) Deactivation of guinea pig pulmonary macrophage responses to N-formyl-methionyl-leucyl-phenylalanine: chemotaxis, superoxide generation, and binding. J Immunol 134:1823–1826

Donabedian H, Gallin JI (1981) Deactivation of human neutrophil chemotaxis by chemoattractants: effect of receptors for the chemotactic factor f-Met-Leu-Phe. J Immunol 127:839–844

English D, Roloff JS, Lukens JN (1981) Regulation of human polymorphonuclear leukocyte superoxide release by cellular responses to chemotactic peptides. J Immunol 126:165–171

Falk W, Leonard EJ (1980) Human monocyte chemotaxis: migrating cells are a subpopulation with multiple chemotaxin specificities on each cell. Infect Immun 29:953–959

Fantone JC, Ward PA (1982) Role of oxygen-derived free radicals and metabolites in leukocyte-dependent inflammatory reactions. Am J Pathol 107:397–418

Farnam J, Grant JA, Lett-Brown MA, Hunt C, Thueson DO, Giclas PC (1985) Complement- and IgE-mediated release of histamine from basophils in vitro. V. Differential effects of drugs modulating arachidonic acid metabolism. J Immunol 134:541–547

Fehr J, Huber A (1984) Complement-induced granulocyte adhesion and aggregation are mediated by different factors: evidence for non-equivalence of the two cell functions. Immunology 53:583–593

Fehr J, Jacob HS (1977) In vitro granulocyte adherence and in vivo margination: two associated complement-dependent functions. Studies based on the acute neutropenia of filtration leukophoresis. J Exp Med 146:641–652

Ferluga J, Schorlemmer HU, Baptista LC, Allison AC (1976) Cytolytic effects of the complement cleavage product, C3a. Br J Cancer 34:626–634

Ferluga J, Schorlemmer HU, Baptista LC, Allison AC (1978) Production of the complement cleavage product, C3a, by activated macrophages and its tumorolytic effects. Clin Exp Immunol 31:512–517

Fernandez HN, Henson PM, Otani A, Hugli TE (1978) Chemotactic response to human C3a and C5a anaphylatoxins. I. Evaluation of C3a and C5a leukotaxis in vitro and under stimulated in vivo conditions. J Immunol 128:109–115
Flohe L, Giertz H, Beckmann R (1985) Free radical scavengers as anti-inflammatory drugs? In: Bonta IL, Bray MA, Parnham MJ (eds) The pharmacology of inflammation. Elsevier, Amsterdam, pp 255–281 (Handbook of inflammation, vol 5)
Forrest MJ, Peck MJ, Williams TJ (1981) Is increased vascular permeability induced by C5a dependent on the generation of lipoxygenase products? Br J Pharmacol 74:921P–922P
Fowler AA, Fisher BJ, Centor RM, Carchman RA (1984) Development of the adult respiratory distress syndrome: progressive alteration of neutrophil chemotactic and secretory processes. Am J Pathol 116:427–435
Fricke D, Damerau B, Vogt W (1985a) Adhesion of guinea pig polymorphonuclear leukocytes to autologous aortic strips: influence of chemotactic factors and of pharmacological agents which affect arachidonic acid metabolism. Int Arch Allergy Appl Immunol 78:429–437
Fricke D, Haller Y, Grünefeld E, Damerau B (1985b) Influence of a prostacyclin analogue and of two drugs affecting arachidonic acid metabolism on leukocyte adhesion. Naunyn Schmiedebergs Arch Pharmacol [Suppl] 329:R 72
Friedberg KD, Engelhardt G, Meineke F (1964) Untersuchungen über die Anaphylatoxin-Tachyphylaxie und über ihre Bedeutung für den Ablauf echter anaphylaktischer Reaktionen. Int Arch Allergy Appl Immunol 25:154–181
Friedberger E (1910) Untersuchungen über Eiweißanaphylaxie. IV. Mitteilung. Z Immunitätsforsch Exp Ther 4:636–689
Gallin JI, Wright DG, Schiffmann E (1978) Role of secretory events in modulating human neutrophil chemotaxis. J Clin Invest 62:1364–1374
Gee MH, Perkowski SZ, Tahamont MV, Flynn JT (1985) Arachidonate cyclooxygenase metabolites as mediators of complement-initiated lung injury. Fed Proc 44:46–52
Gerard C, Hugli TE (1981) Identification of classical anaphylatoxin as the desArg form of the C5a molecule: evidence of a modulator role for the oligosaccharide unit in human desArg^{74}-C5a. Proc Natl Acad Sci USA 78:1833–1837
Gerard C, Chenoweth DE, Hugli TE (1981) Response of human neutrophils to C5a: a role for the oligosaccharide moiety of human C5a-desArg-74 but not of C5a in biologic activity. J Immunol 127:1978–1982
Gerard C, McPhail LC, Marfat A, Stimler-Gerard NP, Bass DA, McCall CE (1986) Role of protein kinases in stimulation of human polymorphonuclear leukocyte oxidative metabolism by various agonists. Differential effects of a novel protein kinase inhibitor. J Clin Invest 77:61–65
Gervasoni JE, Conrad DH, Hugli TE, Schwartz LB, Ruddy S (1986) Degradation of human anaphylatoxin C3a by rat peritoneal mast cells: a role for the secretory granule enzyme chymase and heparin proteoglycan. J Immunol 136:285–292
Giertz H, Hahn F (1966) Makromolekulare Histaminliberatoren. In: Rocha e Silva M (ed) Histamine and anti-histaminics. Springer, Berlin Heidelberg New York, pp 482–568 (Handbook of experimental pharmacology, vol 18/1)
Gimbrone MA, Brock AF, Schafer AI (1984) Leukotriene B_4 stimulates polymorphonuclear leukocyte adhesion to cultured endothelial cells. J Clin Invest 74:1552–1555
Glovsky MM, Hugli TE, Hartman CT, Ghekiere L (1978) Possible role of C3a in human diesease. In: Opferkuch W, Rother K, Schultz DR (eds) Clinical aspects of the complement system. Thieme, Stuttgart, pp 138–144
Glovsky MM, Hugli TE, Ishizaka T, Lichtenstein LM, Erickson BW (1979) Anaphylatoxin-induced histamine release with human leukocytes. Studies of C3a leukocyte binding and histamine release. J Clin Invest 64:804–811

Goetzl EJ (1980) A role for endogenous mono-hydroxyeicosatetraenoic acids (HETEs) in the regulation of human neutrophil migration. Immunology 40: 709–719

Goldstein IM (1979) Endogenous regulation of complement (C5)-derived chemotactic activity: fine-tuning of inflammation: J Lab Clin Med 93:13–16

Goldstein IM, Hoffstein ST, Gallin J, Weissmann G (1973) Mechanisms of lysosomal enzyme release from human leukocytes: microtubule assembly and membrane fusion induced by a component of complement. Proc Natl Acad Sci USA 70: 2916–2920

Goldstein IM, Hoffstein ST, Weissmann G (1975a) Influence of divalent cations upon complement-mediated enzyme release from human polymorphonuclear leukocytes. J Immunol 115:665–669

Goldstein IM, Roos D, Kaplan HB, Weissmann G (1975b) Complement and immunoglobulins stimulate superoxide production by human leukocytes independently of phagocytosis. J Clin Invest 56:1155–1163

Goodman MG, Weigle WO, Hugli TE (1980) Inability of the C3a anaphylatoxin to promote cellular lysis. Nature 283:78–80

Goodman MG, Chenoweth DE, Weigle WO (1982a) Potentiation of the primary humoral immune response in vitro by C5a anaphylatoxin. J Immunol 129: 70–75

Goodman MG, Chenoweth DE, Weigle WO (1982b) Induction of interleukin 1 secretion and enhancement of humoral immunity by binding of human C5a to macrophage surface C5a receptors. J Exp Med 156:912–917

Gorski JP, Hugli TE, Müller-Eberhard HJ (1979) C4a: the third anaphylatoxin of the human complement system. Proc Natl Acad Sci USA 76:5299–5302

Grant JA, Settle L, Whorton EB, Dupree E (1976) Complement-mediated release of histamine from human basophils. II. Biochemical characterization of the reaction. J Immunol 117:450–456

Grant JA, Dupree E, Thueson DO (1977) Complement-mediated release of histamine from human basophils. III. Possible regulatory role of microtubules and microfilaments. J Allergy Clin Immunol 60:306–311

Greef K, Benfey BG, Bokelmann A (1959) Anaphylaktische Reaktionen am isolierten Herzvorhofpräparat des Meerschweinchens und ihre Beeinflussung durch Antihistaminica, BOL, Dihydroergotamin und Reserpin. Naunyn Schmiedebergs Arch Exp Pathol Pharmakol 236:421–434

Grossklaus C, Damerau B, Lemgo E, Vogt W (1976) Induction of platelet aggregation by the complement-derived peptides C3a and C5a. Naunyn Schmiedebergs Arch Pharmacol 295:71–76

Hachfeld del Balzo U, Levi R, Polley MJ (1985) Cardiac dysfunction caused by purified human C3a anaphylatoxin. Proc Natl Acad Sci USA 82:886–890

Hakansson L, Venge P (1985) The chemotactic response of granulocytes to the low molecular weight chemottractants f-MLP, C5f, and LTB4 is dependent on chemokinetic factors. J Leukocyte Biol 38:521–530

Hammerschmidt DE, Craddock PR, McCullogh J, Kronenberg RS, Dalmasso AP, Jacob HS (1978a) Complement activation and pulmonary leukostasis during nylon fiber filtration leukapheresis. Blood 51:721–729

Hammerschmidt DE, White JG, Craddock PR, Jacob HS (1979) Corticosteroids inhibit complement-induced granulocyte aggregation. A possible mechanism for their efficacy in shock states. J Clin Invest 63:798–803

Hammerschmidt DE, Greenberg CS, Yamada O, Craddock PR, Jacob HS (1981a) Cholesterol and atheroma lipids activate complement and stimulate granulocytes. A possible mechanism for amplification of ischemic injury in atherosclerotic states. J Lab Clin Med 98:68–77

Hammerschmidt DE, Harris PD, Wayland JH, Craddock PR, Jacob HS (1981b) Complement-induced granulocyte aggregation in vivo. Am J Pathol 102:146–150

Hartman CT, Glovsky MM (1981) Complement activation requirements for histamine release from human leukocytes: influence of purified $C3a_{hu}$ and $C5a_{hu}$ on histamine release. Int Arch Allergy Appl Immunol 66:274–281
Hartung HP, Hadding U (1983) Complement components in relation to macrophage function. Agents Actions 13:415–428
Hartung HP, Bitter-Suermann D, Hadding U (1983) Induction of thromboxane release from macrophages by anaphylatoxic peptide C3a of complement and synthetic hexapeptide C3a 72–77. J Immunol 130:1345–1349
Heideman M, Hugli TE (1984) Anaphylatoxin generation in multisystem organ failure. J Trauma 24:1038–1043
Henson PM, Zanolari B, Schwartzman NA, Hong SR (1978) Intracellular control of human neutrophil secretion. I. C5a-induced stimulus-specific desensitization and the effects of cytochalasin B. J Immunol 121:851–855
Henson PM, McCarthy K, Larsen GL, Webster RO, Giclas PC, Dreisin RB, King TE, Shaw JO (1979) Complement fragments, alveolar macrophages, and alveolitis. Am J Pathol 97:93–105
Henson PM, Schwartzman NA, Zanolari B (1981a) Intracellular control of human neutrophil secretion. II. Stimulus specificity of desensitization induced by six different soluble and particulate stimuli. J Immunol 127:754–759
Henson PM, Webster RO, Henson JE (1981b) Neutrophil and monocyte activation and secretion: role of surfaces in inflammatory reactions and in vitro. In: Dingle JT, Gordon JL (eds) Cellular interactions. Elsevier/North Holland, Amsterdam, pp 43–56 (Research monographs in cell and tissue physiology, vol 6)
Hirani S, Fair DS, Papin RA, Sundsmo JS (1985) Leukocyte complement: interleukin-like properties of factor Bb. Cell Immunol 92:235–246
Hook WA, Siraganian RP (1977) Complement-induced histamine release from human basophils. III. Effect of pharmacological agents. J Immunol 118: 679–684
Hook WA, Siraganian RP, Wahl SM (1975) Complement-induced histamine release from human basophils. I. Generation of activity in human serum. J Immunol 114:1185–1190
Hoover RL, Folger R, Hearing WA, Ware BR, Karnovsky MJ (1980) Adhesion of leukocytes to endothelium: roles of divalent cations, surface charge, chemotactic agents and substrate. J Cell Sci 45:73–86
Hoover RL, Karnovsky MJ, Austen KF, Corey EJ, Lewis RA (1984) Leukotriene B_4 action on endothelium mediates augmented neutrophil/endothelial adhesion. Proc Natl Acad Sci USA 81:2191–2193
Huey R, Bloor CM, Kawahara MS, Hugli TE (1983) Potentiation of the anaphylatoxins in vivo using an inhibitor of serum carboxypeptidase N (SCPN). I. Lethality and pathologic effects on pulmonary tissue. Am J Pathol 112:48–60
Huey R, Bloor CM, Hugli TE (1984a) Effects of human anaphylatoxins on guinea pig atria. Immunopharmacology 8:147–154
Huey R, Erickson BW, Bloor CM, Hugli TE (1984b) Contraction of guinea pig lung by synthetic oligopeptides related to human C3a. Immunopharmacology 8: 37–45
Hugli TE (1978) Chemical aspects of serum anaphylatoxins. Contemp Top Mol Immunol 7:181–214
Hugli TE (1984) Structure and function of the anaphylatoxins. Springer Semin Immunopathol 7:193–219
Hugli TE (1986) Complement factors and leukocyte activation. J Cell Biochem [Suppl] 10A:242
Hugli TE, Erickson BE (1977) Synthetic peptides with the biological activities and specificity of human C3a anaphylatoxin. Proc Natl Acad Sci USA 74: 1826–1830

Hugli TE, Marceau F (1985) Effects of the C5a anaphylatoxin and its relationship to cyclooxygenase metabolites in rabbit vascular strips. Br J Pharmacol 84: 725–733

Hugli TE, Morgan EL (1984) Mechanisms of leukocyte regulation by complement-derived factors. Contemp Top Immunobiol 14:109–153

Hugli TE, Müller-Eberhard HJ (1978) Anaphylatoxins: C3a and C5a. Adv Immunol 26:1–53

Hugli TE, Gerard C, Kawahara M, Scheetz ME, Barton R, Briggs S, Koppel G, Russell S (1981a) Isolation of three separate anaphylatoxins from complement-activated human serum. Mol Cell Biochem 41:59–66

Hugli TE, Stimler NP, Gerard C, Moon KE (1981b) Possible role of serum anaphylatoxins in hypersensitivity reactions. Int Arch Allergy Appl Immunol [Suppl 1] 66:113–120

Issekutz AC, Movat KW, Movat HZ (1980) Enhanced vascular permeability and haemorrhage-inducing activity of rabbit C5a-desArg: probable role of polymorphonuclear leukocyte lysosomes. Clin Exp Immunol 41:512–520

Issekutz AC, Rochon Y, Ripley M (1985) Synergistic inhibition of complement induced granulocyte margination by BW 755C and calcium channel blockers. Int J Immunopharmacol 7:791–800

Jacob HS (1981) The role of activated complement and granulocytes in shock states and myocardial infarction. J Lab Clin Med 98:645–653

Jacob HS, Craddock PR, Hammerschmidt DE, Moldow CF (1980) Complement-induced granulocyte aggregation. An unsuspected mechanism of disease. N Engl J Med 302:789–794

Jensen JA (1967) Anaphylatoxin and its relation to the complement system. Science 155:1122–1123

Jensen JA, Snyderman R, Mergenhagen SE (1969) Chemotactic activity, a property of guinea-pig C'5-anaphylatoxin: In: Movat HZ (ed) Cellular and humoral mechanisms in anaphylaxis and allergy. Karger, Basel, pp 265–278

Johnson AR, Hugli TE, Müller-Eberhard HJ (1975) Release of histamine from rat mast cells by complement peptides C3a and C5a. Immunology 28:1067–1080

Johnson RJ, Chenoweth DE (1985) Labeling the granulocyte C5a receptor with a unique photoreactive probe. J Biol Chem 260:7161–7164

Kalyanaraman B, Sohnle PG (1985) Generation of free radical intermediates from foreign compounds by neutrophil-derived oxidants. J Clin Invest 75:1618–1622

Kay AB, Shin HS, Austen KF (1973) Selective attraction of eosinophils and synergism between eosinophil chemotactic factor of anaphylaxis (ECF-A) and a fragment cleaved from the fifth component of complement (C5a). Immunology 24: 969–976

Keller HU, Sorkin E (1965) Studies on chemotaxis. II. The significance of normal sera for chemotaxis induced by various agents. Immunology 9:441–447

Keller HU, Wissler JH, Hess MW, Cottier H (1977) Relation between stimulus intensity and neutrophil chemotactic response. Experientia 33:534–536

Keller HU, Wissler JH, Hess MW, Cottier H (1978) Distinct chemokinetic and chemotactic responses in neutrophil granolocytes. Eur J Immunol 8:1–7

Keller HU, Wissler JH, Damerau B (1981) Diverging effects of chemotactic serum peptides and synthetic f-Met-Leu-Phe on neutrophil locomotion and adhesion. Immunology 42:379–383

Kleine I, Poppe B, Vogt W (1970) Functional identity of anaphylatoxin preparations obtained from different sources and by different activation procedures. 1. Pharmacological experiments. Eur J Pharmacol 10:398–403

Kreutzer DL, O'Flaherty JT, Orr W, Showell HJ, Ward PA, Becker EL (1978) Quantitative comparisons of various biological responses of neutrophils to different active and inactive chemotactic factors. Immunopharmacology 1:39–47

Kreutzer DL, Desai U, Orr W, Showell H, Ward PA (1979) Induction of acute inflammatory reactions in lung following intrapulmonary instillation of preformed chemotactic peptides and purified complement components. Chest 75:259–262
Kreuzpaintner G, Damerau B, Zimmermann B, Plummer TH, Brade V (1986) Inactivation of C3a by a monocarboxypeptidase present in culture supernatants of stimulated guinea pig peritoneal macrophages. J Immunol 136:3384–3389
Kunkel SL, Kaercher K, Plewa M, Fantone JC, Ward PA (1982) Production of cyclooxygenase products and superoxide anion by macrophages in response to chemotactic factors. Prostaglandins 24:789–799
Lam WC, Delikatny EJ, Orr FW, Wass J, Varani J, Ward PA (1981) The chemotactic response of tumor cells. A model for cancer metastasis. Am J Pathol 104: 69–76
Lappin D, Damerau B, Whaley K (1983) Anaphylatoxins inhibit C2 production. Clin Exp Immunol 54:455–460
Larsen GL, McCarthy K, Webster RO, Henson J, Henson PM (1980) A differential effect of C5a and C5a-desArg in the induction of pulmonary inflammation. Am J Pathol 100:179–192
Larsen GL, Webster RO, Worthen GS, Gumbay RS, Henson PM (1985) Additive effect of intravascular complement activation and brief episodes of hypoxia in producing increased permeability in the rabbit lung. J Clin Invest 75:902–910
Laude M, Haeffner-Cavaillon N, Fischer E, Cavaillon JM, Kazatchkine MD (1985) Human C3 and C3a stimulate IL-1 secretion by cultured human monocytes. Complement 2:48
Lepow IH, Dias da Silva W, Patrick RA (1969) Biologically active cleavage products of components of complement. In: Movat HZ (ed) Cellular and humoral mechanism in anaphylaxis and allergy. Karger, Basel, pp 237–252
Lett-Brown MA, Boetcher DA, Leonard EJ (1976) Chemotactic responses of normal human basophils to C5a and to lymphocyte-derived chemotactic factor. J Immunol 117:246–252
Levi R, Chenouda AA, Trzeciakowski JP, Guo ZG, Aaronson LM, Luskind RD, Lee CH (1982) Dysrhythmias caused by histamine release in guinea pig and human hearts. Klin Wochenschr 60:965–971
Lunec J, Blake DR, McCleary SJ, Brailsford S, Bacon PA (1985) Self-perpetuating mechanisms of immunoglobulin G aggregation in rheumatoid inflammation. J Clin Invest 76:2084–2090
Magro AM (1982) Effect of inhibitors of arachidonic acid metabolism upon IgE and non-IgE-mediated histamine release. Int J Immunopharmacol 4:15–20
Maly FE, Kapp A, Rother U (1983) C5a-induced chemiluminescence of human granulocytes and its amplification by a serum factor. Immunobiology 164: 90–97
Manderino GL, Suarez AF, Kunkel SL, Ward PA, Hirata AA, Showell HJ (1982) Purification of human C5a desArg by immunoadsorbent and molecular sieve chromatography. J Immunol Methods 53:41–50
Marceau F, Hugli TE (1984) Effect of C3a and C5a anaphylatoxins on guinea pig isolated blood vessels. J Pharmacol Exp Ther 230:749–754
Marder SR, Chenoweth DE, Goldstein IM, Perez HD (1985) Chemotactic responses of human peripheral blood monocytes to the complement-derived peptides C5a and C5a desArg. J Immunol 134:3325–3331
Marom Z, Shelhamer J, Berger M, Frank M, Kaliner M (1985) Anaphylatoxin C3a enhances mucous glycoprotein release from human airways in vitro. J Exp Med 161:657–668
McCall CE, de Chatelet LR, Brown D, Lachmann P (1974) New biological activity following intravascular activation of the complement cascade. Nature 249: 841–843

McCarthy K, Henson PM (1979) Induction of lysosomal enzyme secretion by alveolar macrophages in response to the purified complement fragments C5a and C5a-desArg. J Immunol 123:2511–2517

McDonald JWD, Ali M, Morgan E, Townsend ER, Cooper JD (1983) Thromboxane synthesis by sources other than platelets in association with complement-induced pulmonary leukostasis and pulmonary hypertension in sheep. Circ Res 52:1–6

McPhail LC, Snyderman R (1983) Activation of the respiratory burst enzyme in human polymorphonuclear leukocytes by chemoattractants and other soluble stimuli. Evidence that the same oxidase is activated by different transductional mechanisms. J Clin Invest 72:192–200

Meuer S, Ecker U, Hadding U, Bitter-Suermann D (1981a) Platelet-serotonin release by C3a and C5a: two independent pathways of activation. J Immunol 126: 1506–1509

Meuer S, Hadding U, Andreatta R, Bitter-Suermann D (1981b) Synthetic C3a analogs as specific inhibitors of C3a activity. Immunopharmacology 3:275–280

Meuer S, Hugli TE, Andreatta RH, Hadding U, Bitter-Suermann D (1981c) Comparative study on biological activities of various anaphylatoxins (C4a, C3a, C5a). Investigations on their ability to induce platelet secretion. Inflammation 5: 263–273

Meuer S, Zanker B, Hadding U, Bitter-Suermann D (1982) Low zone desensitization: a stimulus-specific control mechanism of cell response. Investigations on anaphylatoxin-induced platelet secretion. J Exp Med 155:698–710

Morgan EL, Weigle WO, Hugli TE (1982) Anaphylatoxin-mediated regulation of the immune response. I. C3a-mediated suppression of human and murine humoral immune responses. J Exp Med 155:1412–1426

Morgan EL, Thoman ML, Weigle WO, Hugli TE (1983a) Anaphylatoxin-mediated regulation of the immune response. II. C5a-mediated enhancement of human humoral and T cell-mediated immune responses. J Immunol 130:1257–1261

Morgan EL, Weigle WO, Erickson BW, Fok K-F, Hugli TE (1983b) Suppression of humoral immune responses by synthetic C3a peptides. J Immunol 131: 2258–2261

Morgan EL, Thoman ML, Hobbs MV, Weigle WO, Hugli TE (1985a) Human C3a-mediated suppression of the immune response. II. Suppression of human in vitro polyclonal antibody responses occurs through generation of nonspecific $OKT8^+$ suppressor T cells. Clin Immunol Immunopathol 37:114–123

Morgan EL, Thoman ML, Weigle WO, Hugli TE (1985b) Human C3a-mediated suppression of the immune response. I. Suppression of murine in vitro antibody responses occurs through the generation of nonspecific $Lyt\text{-}2^+$ suppressor T cell. J Immunol 134:51–57

Movat HZ, Rettl C, Burrowes CE, Johnston MG (1984) The in vivo effect of leukotriene B_4 on polymorphonuclear leukocytes and the microcirculation. Comparison with activated complement (C5a-desArg) and enhancement by prostaglandin E_2. Am J Pathol 115:233–244

Müller-Eberhard HJ, Bokisch VA, Budzko DB (1971) Studies on human anaphylatoxins and of their physiological control mechanisms. In: Miescher PA (ed) Immunopathology. Karger, Basel, pp 191–200

Muhfelder TW, Niemetz J, Kreutzer D, Beebe D, Ward PA, Rosenfeld SI (1979) C5 chemotactic fragment induces leukocyte production of tissue factor activity. A link between complement and coagulation. J Clin Invest 63:147–150

Musson RA, Becker EL (1976) The inhibitory effect of chemotactic factors on erythrophagocytosis by human neutrophils. J Immunol 117:433–439

Naef A, Damerau B, Keller HU (1982) Cytotaxin-induced cAMP peak in granulocytes: its relationship to crawling movements, chemokinesis and chemotaxis. Biochem Pharmacol 31:1573–1577

Naef A, Damerau B, Keller HU (1984) Relationship between the transient cAMP increase, exocytosis from specific and azurophil granules and chemotaxis in neutrophil granulocytes. Agents Actions 14:63–71

O'Flaherty JT, Craddock PR, Jacob HS (1977a) Mechanisms of anti-complementary activity of corticosteroids in vivo: possible relevance in endotoxin shock. Proc Soc Exp Biol Med 154:206–209

O'Flaherty JT, Kreutzer DL, Showell HJ, Ward PA (1977b) Influence of inhibitors of cellular function on chemotactic factor-induced neutrophil aggregation. J Immunol 119:1751–1756

O'Flaherty JT, Kreutzer DL, Ward PA (1977c) Neutrophil aggregation and swelling induced by chemotactic agents. J Immunol 119:232–239

O'Flaherty JT, Showell HJ, Ward PA (1977d) Neutropenia induced by systemic infusion of chemotactic factors. J Immunol 118:1586–1589

O'Flaherty JT, Craddock PR, Jacob HS (1978a) Effect of intravascular complement activation on granulocyte adhesiveness and distribution. Blood 51:731–739

O'Flaherty JT, Kreutzer DL, Ward PA (1978b) Chemotaxis factor influences on the aggregation, swelling, and foreign surface adhesiveness of human leukocytes. Am J Pathol 90:537–550

O'Flaherty JT, Kreutzer DL, Showell HJ, Vitkauskas G, Becker EL, Ward PA (1979a) Selective neutrophil desensitization to chemotactic factors. J Cell Biol 80: 564–572

O'Flaherty JT, Showell HS, Becker EL, Ward PA (1979b) Neutrophil aggregation: effect of the time and sequence of addition of calcium, magnesium and chemotactic factor. J Recticuloendothel Soc 25:29–38

O'Flaherty JT, Showell HJ, Becker EL, Ward PA (1979c) Neutrophil aggregation and degranulation: effect of arachidonic acid. Am J Pathol 95:433–444

O'Flaherty JT, Showell HJ, Ward PA, Becker EL (1979d) A possible role of arachidonic acid in human neutrophil aggregation and degranulation. Am J Pathol 96:799–809

O'Flaherty JT, Hammett MJ, Shewmake TB, Wykle RL, Love SH, McCall CE, Thomas MJ (1981a) Evidence for 5,12-dihydroxy-6,8,10,14-eicosatetraenoate as a mediator of human neutrophil aggregation. Biochem Biophys Res Commun 103:552–558

O'Flaherty JT, Lees CJ, Miller CH, McCall CE, Lewis JC, Love SH, Wykle RL (1981b) Selective desensitization of neutrophils: further studies with 1-O-alkyl-sn-glycero-3-phosphocholine analogues. J Immunol 127:731–737

O'Flaherty JT, Wykle RL, McCall CE, Shewmake TB, Lees CJ, Thomas M (1981c) Desensitization of human neutrophil degranulation response: studies with 5,12-dihydroxy-6,8,10,14-eicosatetraenoic acid. Biochem Biophys Res Commun 101:1290–1296

O'Flaherty JT, Schmitt JD, McCall CE, Wykle RL (1984) Diacylglycerols enhance human neutrophil degranulation responses: relevancy to a multiple mediator hypothesis of cell function. Biochem Biophys Res Commun 123:64–70

Ogawa H, Kunkel SL, Fantone JC, Ward PA (1981) Comparative study of eosinophil and neutrophil chemotaxis and enzyme release. Am J Pathol 105:149–155

Orr W, Varani J, Ward PA (1978) Characteristics of chemotactic response of neoplastic cells to a factor derived from 5th component of complement. Am J Pathol 93:405–422

Orr W, Phan SH, Varani J, Ward PA, Kreutzer DL, Webster RO, Henson PM (1979a) Chemotactic factor for tumor cells derived from the C5a fragment of complement component C5. Proc Natl Acad Sci USA 76:1986–1989

Orr W, Varani J, Kreutzer DL, Senior RM, Ward PA (1979b) Digestion of the fifth component of complement by leukocyte enzymes. Sequential generation of chemotactic activities for leukocytes and tumor cells. Am J Pathol 94:75–84

Orr FW, Mokashi S, Delikatny J (1982) Generation of complement-derived chemotactic factor for tumor cells in experimentally induced peritoneal exudates and its effect on the local metastasis of circulating tumor cells. Am J Pathol 108: 112–118

Osler AG, Randall HG, Hill BM, Ovary Z (1959) Studies on the mechanism of hypersensitivity phenomena. III. The participation of complement in the formation of anaphylatoxin. J Exp Med 110:311–339

Otte G, Dameran B (1986) Influence of complement peptides on PMN/endothelial adhesion (Abstr) Sixth International Congress of Immunology, Toronto

Otte G, Dameran B, Vogt W (1985) The complement peptides C3a, C5a and their desArg-derivatives enhance adhesion of guinea-pig polymorphonuclear leukocytes (PMN) to autologous aortic strips. Immunobiology 170:70–71

Parkos CA, Cochrane CG, Schmitt M, Jesaitis AJ (1985) Regulation of the oxidative response of human granulocytes to chemoattractants. No evidence for stimulated traffic of redox enzymes between endo and plasma membranes. J Biol Chem 260:6541–6547

Patrick RA, Hollers JC, Liu DY, Heerdt Giese B, Smith CW (1980) Effects of human complement component 1 inactivator on neutrophil chemotaxis and chemotactic deactivation. Infect Immun 28:700–707

Pavek K, Piper PJ, Smedegard G (1979) Anaphylatoxin-induced shock and two patterns of anaphylactic shock: hemodynamics and mediators. Acta Physiol Scand 105:393–403

Payan DG, Trentham DE, Goetzl EJ (1982) Modulation of human lymphocyte function by C3a and C3a(70–77). J Exp Med 156:756–765

Pelayo JC, Chenoweth DE, Hugli TE, Wilson CB, Blantz RC (1986) Effects of the anaphylatoxin, C5a, on renal and glomerular hemodynamics in the rat. Kidney 30:62–67

Perez HD, Hooper C (1986) A specific inhibitor of complement (C5)-derived chemotactic activity in patients with lupus is immunologically related to the Bb fragment of factor B. Clin Res 34:621A

Perez HD, Lipton M, Goldstein IM (1978) A specific inhibitor of complement (C5)-derived chemotactic activity in serum from patients with systemic lupus erythematosus. J Clin Invest 62:29–38

Perez HD, Goldstein IM, Chernoff D, Webster RO, Henson PM (1980) Chemotactic activity of C5a-desArg: evidence of a requirement for an anionic peptide "helper factor" and inhibition by a cationic protein in serum from patients with systemic lupus erythematosus. Mol Immunol 17:163–169

Perez HD, Goldstein IM, Webster RO, Henson PM (1981) Enhancement of the chemotactic activity of human C5a-desArg by an anionic polypeptide ("cochemotaxin") in normal serum and plasma. J Immunol 126:800–804

Petersson BA, Nilsson A, Stalenheim G (1975) Induction of histamine release and desensitization in human leukocytes. Effect of anaphylatoxin. J Immunol 114:1581–1584

Pohajdak B, Gomez J, Orr FW, Khalil N, Talgoy M, Greenberg AH (1986) Chemotaxis of large granular lymphocytes. J Immunol 136:278–284

Polley MJ, Nachman RL (1983) Human platelet activation by C3a and C3a-desArg. J Exp Med 158:603–615

Rampart M, Williams TJ (1985) (without title) Br J Pharmacol 85:274P

Rampart M, Bult H, Herman AG (1983) Activated complement and anaphylatoxins increase the in vitro production of prostacyclin by rabbit aorta endothelium. Naunyn Schmiedebergs Arch Pharmacol 322:158–165

Rampart M, Jose PJ, Williams TJ (1985) Leucocytes activated by C5a desArg promote endothelial prostacyclin (PGI_2) production via release of oxygen species in vitro. Agents Actions 16:21–22

Regal JF (1982) C5a-induced aortic contraction: effect of an antihistamine and inhibitors of arachidonate metabolism. J Pharmacol Exp Ther 220:102–107

Regal JF (1985) Effect of C5a on isolated guinea pig atria. Immunopharmacology 9:27–31

Regal JF, Pickering RJ (1981) C5a-induced tracheal contract: effect of an SRS-antagonist and inhibitors of arachidonate metabolism. J Immunol 126:313–316

Regal JF, Pickering RJ (1983) C5a and antigen-induced tracheal contractions: effect of a combination of an antihistamine and cyclooxigenase inhibitors: Int J Immunopharmacol 5:71–78

Regal JF, Eastman AY, Pickering RJ (1980) C5a induced tracheal contractions: a histamine independent mechanism. J Immunol 124:2876–2878

Regal JF, Hardy TM, Casey FB, Chakrin LW (1983a) C5a-induced histamine release. Species specificity. Int Arch Allergy Appl Immunol 72:362–365

Regal JF, Hardy TM, Casey FB, Chakrin LW (1983b) Effects of C5a on guinea pig lung: histamine release and mechanism of contraction. Immunopharmacology 5: 315–327

Rivkin I, Rosenblatt J, Becker EL (1975) The role of cyclic AMP in the chemotactic responsiveness and spontaneous motility of rabbit peritoneal neutrophils. The inhibition of neutrophil movement and the elevation of cyclic AMP levels by catecholamines, protaglandins, theophylline and cholera toxin. J Immunol 115:1126–1134

Romualdez AG, Ward PA (1975) A unique complement derived chemotactic factor for tumor cells. Proc Natl Acad Sci USA 72:4128–4132

Romualdez AG, Ward PA, Torikata T (1976) Relationship between the C5 peptides chemotactic for leukocytes and tumor cells. J Immunol 117:1762–1765

Sacks T, Moldow CF, Craddock PR, Bowers TK, Jacob HS (1978) Oxygen radicals mediate endothelial cell damage by complement-stimulated granulocytes. An in vitro model of immune vascular damage. J Clin Invest 61:1161–1167

Scheid CR, Webster RO, Henson PM, Findlay SR (1983) Direct effect of complement factor C5a on the contractile state of isolated smooth muscle cells. J Immunol 130:1997–1999

Schröder JM, Christophers E (1985) Transient absence of C5a-specific neutrophil function in inflammatory disorders of the skin. J Invest Dermatol 85:194–198

Schumacher KA, Classen HG, Hagedorn M, Benner KU, Späth M, Mittermayer C (1974) Effects of anaphylatoxin plasma in cats: hemodynamic changes induced by platelet aggregation. Arzneimittelforsch 24:122–126

Shaw JO, Henson PM, Henson J, Webster RO (1980) Lung inflammation induced by complement-derived chemotactic fragments in the alveolus. Lab Invest 42: 547–558

Shaw JO, Brodersen I, Lyons RM (1982) Extra- and intracellular Ca^{++} requirement for lysosomal enzyme secretion in human neutrophils. Agents Actions 12: 328–332

Shields JM, Haston WS (1985) Behaviour of neutrophil leucocytes in uniform concentrations of chemotactis factors: concentration waves, cell polarity and persistence. J Cell Sci 74:75–93

Shin HS, Snyderman R, Friedman E, Mellors A, Mayer MM (1968) Chemotactic and anaphylatoxic fragment cleaved from the fifth component of guinea pig complement. Science 162:361–363

Shingu M, Nobunaga M (1984) Chemotactic activity generated in human serum from the fifth component of complement by hydrogen peroxide. Am J Pathol 117:201–206

Showell HJ, Becker EL (1976) The effects of external K^+ and Na^+ on the chemotaxis of rabbit peritoneal neutrophils. J Immunol 116:99–105

Showell HJ, Naccache PH, Sha'afi RI, Becker EL (1977) The effect of extracellular K^+, Na^+ and Ca^{++} on lysosomal enzyme secretion from polymorphonuclear leukocytes. J Immunol 119:804–811

Showell HJ, Naccache PH, Sha'fi RI, Becker EL (1980) Inhibition of rabbit neutrophil lysosomal enzyme secretion, non-stimulated and chemotactic factor stimulated locomotion by nordihydroguaiaretic acid. Life Sci 27:421–426
Showell HJ, Glovsky MM, Ward PA (1982a) C3a-induced lysosomal enzyme secretion from human neutrophils. Lack of inhibition by f met-leu-phe antagonists and inhibition by arachidonic acid antagonists. Int Arch Allergy Appl Immunol 67:227–232
Showell HJ, Glovsky MM, Ward PA (1982b) Morphological changes in human polymorphonuclear leukocytes induced by C3a in the presence and absence of cytochalasin B. Int Arch Allergy Appl Immunol 69:62–67
Simchowitz L, Atkinson JP, Spilberg I (1980) Stimulus-specific deactivation of chemotactic factor-induced cyclic AMP response and superoxide generation by human neutrophils. J Clin Invest 66:736–746
Siraganian RP, Hook WA (1976) Complement-induced histamine relase from human basophils. II. Mechanism of the histamine release reaction. J Immunol 116: 639–646
Smedley LA, Tonnesen MG, Sandhaus RA, Haslett C, Guthrie LA, Johnston RB, Henson PM, Worthen GS (1986) Neutrophil-mediated injury to endothelial cells. Enhancement by endotoxin and essential role of neutrophil elastase. J Clin Invest 77:1233–1243
Smith CW, Hollers JC (1980) Motility and adhesiveness in human neutrophils. Redistribution of chemotactic factor-induced adhesion sites. J Clin Invest 65:804–812
Smith CW, Hollers JC, Patrick RA, Hassett C (1979) Motility and adhesiveness in human neutrophils. Effects of chemotactic factors. J Clin Invest 63: 221–229
Snyderman R, Pike MC (1984) Chemoattractant receptors on phagocytic cells. Annu Rev Immunol 2:257–281
Snyderman R, Phillips J, Mergenhagen SE (1970) Polymorphonuclear leukocyte chemotactic activity in rabbit serum and guinea pig serum treated with immune complexes: evidence for C5a as the major chemotactic factor. Infect Immun 1:521–525
Snyderman R, Shin HS, Dannenberg AM (1972) Macrophage proteinase and inflammation: the production of chemotactic activity from the fifth component of complement by macrophage proteinase. J Immunol 109:896–898
Solomkin JS, Jenkins MK, Nelson RD, Chenoweth D, Simmons RL (1981) Neutrophil dysfunction in sepsis. II. Evidence for the role of complement activation products in cellular deactivation. Surgery 90:319–327
Sorgenfrei J, Damerau B, Vogt W (1982) Role of histamine in the spasmogenic effect of the complement peptides C3a and C5a-desArg (classical anaphylatoxin). Agents Actions 12:118–121
Stevens JH, O'Hanley P, Shapiro JM, Mihm FG, Satoh PS, Collins JA, Raffin TA (1986) Effects of anti-C5a antibodies on the adult respiratory distress syndrome in septic primates. J Clin Invest 77:1812–1816
Stimler NP (1984) Spasmogenic activity of C5a-desArg anaphylatoxin on guinea pig lung parenchymal strips: sensitivity of the leukotriene-mediated component to cyclooxygenase inhibitors. Biochem Biophys Res Commun 125:852–858
Stimler NP, Hugli TE, Bloor CM (1980) Pulmonary injury induced by C3a and C5a anaphylatoxins. Am J Pathol 100:327–340
Stilmer NP, Bloor CM, Hugli TE, Wykle RL, McCall, O'Flaherty JT (1981a) Anaphylactic actions of platelet-activating factor. Am J Pathol 105:64–69
Stimler NP, Brocklehurst WE, Bloor CM, Hugli TE (1981b) Anaphylatoxin-mediated contraction of guinea pig lung strips: a nonhistamine tissue response. J Immunol 126:2258–2261
Stimler NP, Bach MK, Bloor CM, Hugli TE (1982) Release of leukotrienes from guinea pig lung stimulated by C5a-desArg anaphylatoxin. J Immunol 128: 2247–2252

Stimler NP, Bloor CM, Hugli TE (1983) C3a-induced contraction of guinea pig lung parenchyma: role of cyclooxygenase metabolites. Immunopharmacology 5:251–257

Sundsmo JS, Götze O (1981) Human monocyte spreading induced by factor Bb of the alternative pathway of complement activation. A possible role for C5 in monocyte spreading. J Exp Med 154:763–777

Till GO, Johnson KJ, Kunkel R, Ward PA (1982) Intravascular activation of complement and acute lung injury. Dependency on neutrophils and toxic oxygen metabolites. J Clin Invest 69:1126–1135

Tonnesen MG, Smedley LA, Henson PM (1984) Neutrophil-endothelial cell interactions. Modulation of neutrophil adhesiveness induced by complement fragments C5a and C5a-desArg and formyl-methionyl-leucyl-phenylalanine in vitro. J Clin Invest 74:1581–1592

Tvedten HW, Till GO, Ward PA (1985) Mediators of lung injury in mice following systemic activation of complement. Am J Pathol 119:92–100

Vallota EH, Müller-Eberhard HJ (1973) Formation of C3a and C5a anaphylatoxins in whole human serum after inhibition of the anaphylatoxin inactivator. J Exp Med 137:1109–1123

Van Epps DE, Garcia ML (1980) Enhancement of neutrophil function as a result of prior exposure to chemotactic factor. J Clin Invest 66:167–175

Vargaftig BB, Lefort J, Wal F, Chignard M, Medeiros MC (1982) Non-steroidal anti-inflammatory drugs if combined with anti-histamine and anti-serotonin agents interfere with the bronchial and platelet effects of "platelet-activating factor" (PAF-acetether). Eur J Pharmacol 82:121–130

Verghese MW, Fox K, McPahil LC, Snyderman R (1985) Chemoattractant-elicited alterations of cAMP levels in human polymorphonuclear leukocytes require a Ca^{+}-dependent mechanism which is independent of transmembrane activation of adenylate cyclase. J Biol Chem 260:6769–6775

Vogt W (1968) Preparation and some properties of anaphylatoxin from hog serum. Biochem Pharmacol 17:727–733

Vogt W (1974) Activation, activities and pharmacologically active products of complement. Pharmacol Rev 26:125–169

Vogt W (1986) Anaphylatoxins: possible roles in disease. Complement 3:177–188

Vogt W, Zeman N, Garbe G (1969) Histaminunabhängige Wirkungen von Anaphylatoxin auf glatte Muskulatur isolierter Organe. Naunyn Schmiedebergs Arch Pharmakol Exp Pathol 262:399–404

Vogt W, Damerau B, Lühmann B, Hesse D, Haller Y (1986) Complement activation in human lymph: modulation by the contact activation system and by leukocytes. Int Arch Allergy Appl Immunol 79:423–433

Ward PA, Cochrane CG, Müller-Eberhard HJ (1965) The role of serum complement in chemotaxis of leukocytes in vitro. J Exp Med 122:327–346

Ward PA, Till GO, Hatherill JR, Annesley TM, Kunkel RG (1985) Systemic complement activation, lung injury, and products of lipid peroxidation. J Clin Invest 76:517–527

Webster RO, Hong SR, Johnston RB, Henson PM (1980a) Biological effects of the human complement fragments C5a and C5a-desArg on neutrophil function. Immunopharmacology 2:201–219

Webster RO, Zanolari B, Henson PM (1980b) Neutrophil chemotaxis in response to surface-bound C5a. Exp Cell Res 129:55–62

Webster RO, Larsen GL, Henson PM (1982) In vivo clearance and tissue distribution of C5a and C5a des arginine complement fragments in rabbits. J Clin Invest 70:1177–1183

Wedmore CV, Williams TJ (1981) Control of vascular permeability by polymorphonuclear leukocytes in inflammation. Nature 289:646–650

Weissmann G (1982) Activation of neutrophils and the lesions of rheumatoid arthritis. J Lab Clin Med 100:322–333

Weissmann G, Smolen JE, Korchak HM (1980) Release of inflammatory mediators from stimulated neutrophils. N Engl J Med 303:27–34

Weissmann G, Smolen J, Korchak H, Hoffstein S (1981) The secretory code of the neutrophil. In: Dingle JT, Gordon JL (eds) Cellular interactions. Elsevier/North-Holland, Amsterdam, pp 15–31 (Research monographs in cell and tissue physiology, vol 6)

Wilkinson PC (1985) A visual study of chemotaxis of human lymphocytes using a collagen-gel assay. J Immunol Methods 76:105–120

Wissler JH, Stecher VJ, Sorkin E (1972) Biochemistry and biology of a leukotactic binary serum peptide system related to anaphylatoxin. Int Arch Allergy Appl Immunol 42:722–747

Wright DG, Gallin JI (1977) A functional differentiation of human neutrophil granules: generation of C5a by a specific (secondary) granule product and inactivation of C5a by azurophil (primary) granule products. J Immunol 119:1068–1076

Wright DG, Gallin JI (1979) Secretory responses of human neutrophils: exocytosis of specific (secondary) granules by human neutrophils during adherence in vitro and during exudation in vivo. J Immunol 123:285–294

Wright CD, Hoffman MD (1986) The protein kinase C inhibitors H-7 and H-9 fail to inhibit human neutrophil activation. Biochem Biophys Res Commun 135:749–755

Wuepper KD, Bokisch VA, Müller-Eberhard HJ, Stoughton RB (1972) Cutaneous responses to human C3a anaphylatoxin in man. Clin Exp Immunol 11:13–20

Yancey KB, Hammer CH, Harvath L, Renfer L, Frank MM, Lawley TJ (1985) Studies on human C5a as a mediator of inflammation in normal human skin. J Clin Invest 75:486–495

Zimmerman GA, Wiseman GA, Hill HR (1985) Human endothelial cells modulate granulocyte adherence and chemotaxis. J Immunol 134:1866–1874

Zimmermann B, Damerau B, Vogt W (1980) Purification and partial amino acid sequence of classical anaphylatoxin from pig serum; identification with desArg-C5a. Hoppe-Seylers Z Physiol Chem 361:915–924

Subject Index

GPSR Compliance

The European Union's (EU) General Product Safety Regulation (GPSR) is a set of rules that requires consumer products to be safe and our obligations to ensure this.

If you have any concerns about our products, you can contact us on ProductSafety@springernature.com

In case Publisher is established outside the EU, the EU authorized representative is:

Springer Nature Customer Service Center GmbH
Europaplatz 3
69115 Heidelberg, Germany

Batch number: 10620881

Printed by Printforce, the Netherlands